Ocean Engineering

Ocean Engineering

Editor
T.V. Ramakrishnan

2007

Gene-Tech Books
New Delhi - 110 002

ISBN 81-89729-11-X

Published by : GENE-TECH BOOKS
4762-63/23, Ansari Road, Darya Ganj,
NEW DELHI - 110 002
Phone: 41562849
e-mail: genetechbooks@yahoo.co.in

Printed at : Chawla Offset Printers
New Delhi – 110 052

PRINTED IN INDIA

PREFACE

Ocean engineering is an important subject which consists three broad categories; port and coastal engineering offshore and continental shelf engineering and deep ocean engineering.

The present book covers the topics like harbor design, fisheries science, oscilab design, marine resources development etc. The book throws light on, dynamic positioning of surface ships and submersibles, motion control at the sea-air interface, etc. The book carries, important topics like sea floor conditions and wave action influences.

The book gives attention at the topics like transport of substances in the sea, the wind generated waves, etc.

I shall regard my efforts amply rewarded if the book is received with the enthusiasm.

I wish to express my sincere thanks to the publishers, especially to Mr. Anil Mittal, for his continued interest in bringing out this edition.

Author

CONTENTS

Chapter One
Introduction

Ocean Engineering is the subject which covers, number of topics like port and coastal engineering, offshore continental shelf engineering, and deep ocean engineering. In the present book particular attention is given to dynamic positioning of surface ships and small submersibles, motion control at the sea-air interface, psychological and physiological aspects of humans during submersed operations.

While studying ocean engineering, the topics like, transport of substances in the sea, the wind generated waves take a prime role.

Problems of diffusion and transport of substances in the sea have received considerable theoretical and experimental attention in recent decades – especially radioactive substances and pollutants. Despite these broad efforts, Okubo concludes that "diffusion is confusion."

Most attempts to find accurate, universal solutions to dispersion formulas have encountered the same difficulties:

- Numerical solutions require simultaneous treatment of many parameters (diffusion, transport, decay, etc.). All of these vary widely in space and time and are not independent.
- Velocity field specification has been incomplete. Fluctuations in transport and turbulence vectors in short time steps were not known. Averaging over any extended period introduces unacceptable errors because the processes are nonlinear.

In spite of the simplifications used to surmount these difficulties, some Eulerian approaches to diffusion problems have given satisfactory results. In most applied problems treating both diffusion and transport, however, a Lagrangian solution is required.

Two recent developments show promise in handling these complicated problems:

- Successful application of hydrodynamical numerical models which include diffusion formulas in finite difference form.
- Use of large computers for synoptic analysis/prediction in the oceans.

The combined effect of these developments permits simultaneous, numerical solution of complex formulations in short time steps. The necessary current vectors and parameter distributions are carried along in a marching process. The solutions are essentially Lagrangian in form although certain Eulerian treatments will be used where possible and necessary.

While we are studying about transport in oceans we must learn about ocean waves. The study of waves begins with the observation of nature. That such is the proper course was already voiced by Leonardo: "If you wish to study water, first introduce experience and then the reason of things." But the observation of waves, particularly those that attain great height or those within a storm-generating area, is fairly unreliable and always difficult if not unmanageable, even with sophisticated instrumentation. Because nature rarely co-operates with the scientist, he must set up in the laboratory models of nature that he can observe in the small but under controllable and reproducible conditions.

This modeling presents difficulties, for not all aspects of nature follow the same scaling law. In addition, some aspects may not be relevant to a line of inquiry. Indeed, the central problem is separating the essential aspects from the trivial ones. This requires insight and, if the results are to be applied in design, keen engineering judgment.

The observation of nature in the large and in the small leads to the formulation of empirical laws of behavior but not directly to an understanding. This can come about only with the development of a coherent theory that provides the reasons for the behavior observed in the laboratory and in nature.

To set up a theory, one formulates a hypothesis, works out the results, and compares these results with those of laboratory experiments or full-scale trials. Disagreement indicates that both hypothesis and measurements are suspect; thus, a restudy is called for. One may claim understanding of the phenomenon only when the results of theory, laboratory experimentation, and full-scale observations agree. One has some way to go before any such statement can be made about ocean waves.

Since the Second World War, observations of ocean waves have become systematic and are carried out by weather ships. Roll compiled the voluminous reports of periodic visual observations. Since the development of a ship-fitted wave gauge, the observations have ceased to be subjective visual estimates. The formulation of empirical laws of the seaway has become the central preoccupation of every ambitious oceanographer.

The classical approach to the solution of wave problems has been by application of perfect-fluid theory. This has led to an understanding of how waves are generated by moving bodies (pistons, flaps, hulls), how they propagate from their origin, how their principal characteristics of height and length change with depth of water, how they are refracted in shoaling water and diffracted by obstacles, and how their height changes with dispersion. But the classical theory provides no answer to how waves are generated by wind and how their height diminishes as the consequence of internal dissipation. In addition, in classical theory, all waves generated under steady condition are of identical characteristics. Although this identical resemblance can be developed in laboratory experiments, its appearance in nature is approached only under certain specific conditions of uncommon occurrence. However, the classical theory does provide certain kinematical and kinetic relations of fundamental importance.

Now let us try to learn about few more things about sea spectra. Purely descriptive representations of the sea spectrum are advantageous for observations in the scientific laboratory where it may not be possible to reproduce faithfully conditions of wind field and fetch. They are also the ones to be employed when discussing the wave-induced forces excited on bodies (such as piles, platforms, buoys, and ships) and the possible response of such bodies to wave action.

Every oceanographer has either devised his own variance spectrum of the seaway or has an obstinate preference for a certain formulation. Thus, a rather larger number of spectra are available, each of which is the result of a curve-fitting process applied to a small sample of particular data. These spectra are advocated by an even larger number of vociferous proponents. Fortunately for the ocean engineering fraternity, the fondly praised and clamorously derogated formulations are not far apart in yielding useful predictions.

The earliest formulation of a sea spectrum is due to Neumann. It is a so-called one-dimensional spectrum (*i.e.*, it relates to the rise and fall of the waves at a point) and reads

$$S_\eta(\omega) = C\omega^{-6} \exp(-2g^2 \omega^{-2} U^{-2})$$

where U is the wind speed and C is an empirical constant. The formula holds for the fully developed sea that corresponds to the wind speed U.

Thus, the formulation is in terms of the cause of the sea and is not a direct description of the seaway itself. It is questionable whether the same law linking U and $S_\eta(\omega)$ (or $S_\eta(T)$) always holds under all meteorological and oceanic conditions.

The Neumann spectrum is mentioned because of its historical importance in complementing the Pierson model of the seaway. The Pierson model required a sea spectrum to be useful, and the Neumann spectrum required a mathematical model for its interpretation.*

In recent years the tendency has been toward a sea spectrum of the form one would have an integral which is unbounded, and

$$(d^2\eta/dt^2)_S = \sqrt{\omega^4 S_\eta/2}$$

$$= \sqrt{\frac{1}{2}\int_0^\infty \omega^4 S_\eta(\omega)d\omega}$$

$$= \sqrt{\frac{1}{2}\int_0^\infty A\omega^{-1} \exp(-B\omega^{-2})\, d\omega}$$

the outcome would be physically inconsistent. From purely mathematical reasoning, the first factor would have to contain the frequency at a negative power greater than the fifth.

Kinsman contends that to debate over the frequency exponents is rather nonsensical since they are obtained as an average of a

---

* The author, not being an oceanographer, is completely unbiased in this regard.

large number of records. The nonlinear interactions between wave components are obscured when the wave spectra are averaged. This is certainly so: the wave spectrum is a valid representation of a linear process but somewhat difficult to employ and interpret when the process is nonlinear. The procedure of using empirical constants obtained by averaging a large number of records results in a pseudolinear system which, although exhibiting a mathematical flaw is nonetheless, of somewhat improved reliability.

Neumann and Pierson suggest that the behavior of the wave spectrum in the high-frequency region is to be attributed to the occurrence of breaking waves. This is, undoubtedly, a good point, but determination of the probability law of wave breaking is likely to prove itself a difficult task.

Oceanic engineering also covers the important topics like, dynamic positioning of surface-ships. The term *dynamic positioning* was coined in 1959 to define a means of holding a ship on the surface of the ocean in a relatively fixed position with respect to the ocean floor without the use of anchors. To the best of my knowledge, the concept of performing this feat with the aid of maneuvering propulsion devices was first suggested when plans were being developed for the Mohole Experimental Drilling Program.

Since that time, the technique has been used on several ships for a variety of purposes, most of them associated with ocean engineering. Means of determining the precise location of the ship, of controlling propulsion units and the performance of thrust-producing devices have improved over the intervening years. The basic design problems, however, have not changed; therefore, I believe that the story of the development of the first ship dynamic-positioning system would be of interest. The lessons learned in planning, designing, building, and operating this system provide a valuable background for those who may design or use such systems.

At the present time there is a strong and persistent debate as to the characteristics of waves developed in an open sea of great depth under given conditions of wind strength, duration, and fetch length. But the much more difficult problem of determining which waves are generated by a wind of given (steady) strength and duration blowing over a shallow body of water of limited fetch is discussed only sporadically.

Progress on the problem has been slow, but such slowness caused little hardship until the recent rapid growth of the offshore industry. The present demand for environmental information for designing offshore structures has raised the problem to a level of primary importance. One should expect that progress will now accelerate.

The method of wave prediction in use at present is a logic which consists in the proper sequencing of empirical formulations. The procedure is largely graphical, hence laborious. Availability of a computer-adapted technique would facilitate the work considerably. In this section, a logic will be outlined that can be coded.

Application of the logic is restricted to cases in which refraction is negligible. Whether or not such a restriction is of serious consequence can only be judged in specific cases. No general statement can be made. However, if the wind blows offshore and the depth of water increases monotonically (or almost so), refraction tends to be of relatively small significance.

There are data on relatively small bodies of shallow water which tend to support the practice of applying to a determination of the characteristics of waves in restricted bodies of water the same empirical relations derived for determining the characteristics of waves in the open sea, provided certain suitable modifications are introduced relative to the measurement of fetch length.

The empirical relations alluded to are those of Sverdrup, Munk, and Bretschneider (SMB). These link significant wave height (H) and wave period (T) to wind speed (U), fetch length (F), water depth (d), and wind duration (t).

To be useful in computer solutions, the relations need be expressed numerically rather than graphically.

To this end, several formulations are available, albeit not all in relative mutual accord. This should not be surprising, since observational data on the environment always seem to exhibit a frightful scatter, particularly when the method of observation is visual rather than instrumental.

Consider also that, in the present instance, certain parameters, such as fetch length and wind duration, which must be noted at the time of mking an observation of wave height and period, represent pure judgements on the part of the observer. Occasionally, even an experienced observer finds it difficult to arrive at a valid assessment, for the environment is not a well-behaved system: winds do not suddenly start from a calm and blow at a steady intensity in a fixed direction; fetch length rarely has a fixed boundary to which measurements can be referred.

The empirical relations are written in terms of:

(a) A wave-height parameter, gH/U^2

(b) A wave-period parameter, gT/U

(c) A fetch-length parameter, gF/U^2

where

H = significant wave height, ft

T = significant wave period, sec

F = fetch length, ft

g = gravitational acceleration, ft sec^{-2}

U = wind speed, ft sec^{-1}

Today the subject ocean engineering has got worldwide importance. People are very much interested in, open sea propagation, hydraulic structures of ice, harbor design, fisheries science, air-sea interface, scientific underwater human activities– and also in many more topics.

Chapter Two
The Harbor Design and the Design of the Oscilab

2.1 The Suitable Harbor Design

A harbor is intended to provide a protective environment for ships. The Harbor design is essentially related to the problem of ship motions.

A freely floating vessel in a wave environment moves in a complicated fashion which, for convenience, is generally resolved into six components: heave, pitch and roll, surge, sway and yaw. Of these motions, the first group of three possesses spring-like restoration forces.

The second group of three motions, surge, sway and yaw, is not subject to such an effect if the vessel is free since no restoring force is present. Mooring the vessel provides the necessary restoration by means of the lines, either by weight or by elasticity so that again resonance is possible. It is such resonance in surge, sway or yaw with which one is most concerned since these are motions in the horizontal plane and thus promise. If large enough, collision with the surroundings. We shall not pursue ship motions any further except to mention that if the period is proper, very small amplitude waves are sufficient to cause damage.

Now it is pertinent to inquire just what the exciting waves are like. The waves arriving from deep water towards the coast represent a spectrum of frequencies from the very short to "sea and swell" caused by wind storms, either local or distant, to very long waves. Sea and swell may be thought to possess periods in the range from 4 to 20 seconds, while very long waves of interest may be typically of 1 to 3-minute periods. These ranges are specified to relate to two major classes of boats. Small pleasure boats of 20 to

50-ft. length are vulnerable to the shorter waves with periods, say, from 4 to 10 seconds. Large ships on the other hand, say, a 40,000-ton moored vessel may be vulnerable to waves in the 1 to 2-minute range.

Different design considerations must, of course, be made for the two types of waves as their behavior in a harbor are quite different. For brevity, we shall refer to the shorter waves as wind waves and to the long waves as "seiche"; although seiche is more properly the oscillation within the harbor caused by long waves.

In general, the statistical distribution of energy as a function of frequency at a given location is not very well known beyond the range of wind waves. It is reasonable to expect that the incident deep water energy spectrum is generally fairly flat between 30 to 200 seconds. This range of wave period triggers seiches in a harbor basin which are a cause of concern for large moored ships. However, due to peculiarities of refraction, reflection, multiple reflection and resonance over the continental slope and continental shelf, the shallow water energy spectrum presents a number of irregularities and peaks.

As a matter of fact, physical behavior of long waves depends upon the bottom topography even in the deepest ocean. Moreover, due to small wave steepness the corresponding energy is not dissipated by breaking. The wave reflects back and forth from one continent to another. A relatively high energy level exists on the continental shelf caused by the reflective and refractive power of the continental slope. (Trapped modes, edge waves and bay resonance are of the same family of phenomena which are of particular significance in the case of long waves.)

In the case where preferential modes of oscillations of embayment and continental shelf match some of the dominant natural oscillations of a harbor basin, large seiche amplitude can be observed.

Analysis of long wave records theoretically permits one to determine the probability of occurrence and intensity of the incident long wave energy. Such information rarely exists due to the length of recording time which would be required to obtain any significant results. Theoretically, a scale model or a numerical investigation which will reproduce the bottom topography up to the toe of the continental shelf should enable one to determine not only the response of a harbor basin but also the peculiarities of incident energy level at the entrance of the harbor. For practical purposes, one may prefer to build two scale models at two different scales. The first one at a small scale embracing a large part of the oceanographic topography which would provide the input for the second model reproducing the harbor and its immediate neighborhood only at a relatively larger scale.

In fact, such a theoretically ideal method is rarely accepted in engineering practice for obvious reasons of economy. The problem of seiche is simply solved by comparing various harbor layouts independently of the fact that some modes of oscillation may seldom be excited and others may have a very high probability of occurrence due to shelf and bay reinforcement.

Prior to any investigation on wave agitation in a harbor, one must have a reasonably accurate knowledge of the origin of the phenomena. For example, the scale model as well as the mathematical model will have to be completely different where one assumes that the seiche motion is due to long waves generated by atmospheric disturbances or is due to "surf beats." In the scale model, one has to generate long adjustable periodic waves while in the mathematical model, it may be necessary to reproduce short modulated wave trains of the wind wave range. Also, the scale model may be amenable to satisfactory linearized mathematical modeling. Any harbor can be considered as a resonant box, much like an organ pipe; it possesses natural modes of oscillation which, if excited by waves of corresponding frequency, will be induced to resonate.

The essential mechanism by which resonance is affected is partial reflection by boundaries. Consider a rectangular harbor of constant depth. An entering wave travels to the rear of the harbor where it is reflected by the wall and proceeds back toward the entrance. Upon reaching the entrance, part of the wave is reflected by the sudden enlargement while the rest escapes seaward. The reflected component remains within the harbor and repeats the cycle. Now if the frequency of arrival of waves is just right, this reflected wave combines with a new incident wave and amplification begins. The wave amplitude within the harbor continues to grow until the energy lost seaward at the entrance and frictional or breaking dissipation equals the arriving wave energy.

In the case of wind waves, a great deal of protection is provided by simply preventing their arrival within the harbor through the construction of breakwaters in such a way that the arriving waves are reflected or largely absorbed. Design may be optimized to largely prevent the arrival of wind wave energy by a suitable choice of breakwater layout.

Of course, there must exist an entrance to the harbor and through this some energy will pass. This energy, as well as any surviving through the breakwater, can be effectively absorbed; for example, natural beaches are excellent wave absorbers because of energy dissipation due to wave breaking. Wave traps in rubble mounds are the most common and practical solution.

It is more difficult to prevent the arrival of seiche within a harbor. The origin of seiche may be of several types. Seismic disturbance, large scale atmospheric phenomena, trapping of energy on the continental shelf, nonlinear interaction between wind waves corresponding to a shift of the energy level from high frequency waves to low frequency waves, surf beats, currents past the harbor entrance (generating alternating Karman vortices), even ship transit in and out of the harbor may induce seiche oscillation.

It is interesting to note that rockfill breakwaters are essentially transparent to very long waves. For example, more wave energy of importance arrives within the Los Angeles Harbor through the breakwater than through the entrance to the harbor. Since breakwaters are transparent in both directions, harbor basins adjacent to breakwaters cannot trap any large amount of energy. So the problem of resonance only exists in inner basins, in which case attempts to limit the level of incident wave energy should be made. For this reason, rather be inclined to make a breakwater as impervious as possible to long waves despite the fact that seiche may now exist in offshore turning basins. If breakwaters can be made essentially impervious, it may suffice to make internal basins, berths and quays "transparent" in order to avoid local amplification. This can be done by building structures on piles through which the energy can easily pass rather than on solid bases between which oscillations may amplify. It is noted that wave absorption of very long waves is relatively inefficient. Long waves of small wave steepness do not break even on relatively gentle beaches and cannot be counted on to hold down the oscillation amplitude.

More often one seeks to design the harbor such that whatever waves are unavoidable do not amplify.

2.2 The Technique of Harbor Design

The most widely adopted technique of harbor design is the scale model study. The simplest way is to construct a model in geometric scale and to determine experimentally its characteristics.

If the seiche motion is due to surf beat, the choice of scale is the same as that which prevails in the case of penetration of wind waves; *i.e.,* the model cannot be distorted and the scale is about 1:150. A smaller scale would cause too much scale effect by viscous and capillary action.The accurate reproduction of the generation process of surf beat reflecting towards the entrance of a harbor may require the reproduction of a large nearby bottom topography section.

For seiche waves, large regions need to be modeled owing to the importance of shelf oscillations etc., and the model may need to cover a whole shelf section. The results of the first scale model can then be used as inputs for the second model which will be built to a larger scale, but covering a smaller, area. If it is possible to treat a harbor separately, the first model can be built to a scale of 1:1000. Such a small scale, however, will usually require a distortion of the vertical scale so that excessively thin water layers will not result. A maximum permissible distortion is obtained by consideration of maintaining similitude of refraction. The general result is that the maximum depth which can be represented in a distorted model is inversely proportional to the square of the distortion; hence, the allowable distortion is limited by the maximum depth of the model and minimum wavelength.

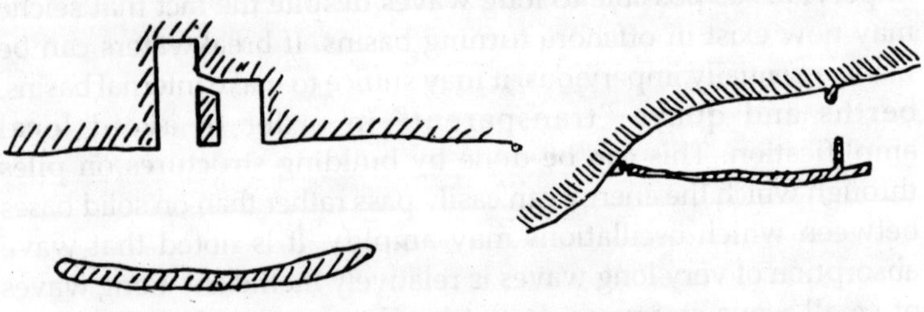

Fig. 2.1 : Providing a "leak," which avoids energy concentration, limits resonance.

Distortion of the depths, however, violates similitude of coastal reflection. It is then possible, and usually desirable, to correct this error by distorting the wave height. Distortion of depth increases reflection; an increase in wave steepness decreases reflection so that some compensation may be achieved.

A further practical problem relates to the satisfaction of boundary conditions. Clearly, a shelf region and harbor are considered as unbounded, being open to the sea. The walls of the test basin and the wave generating paddle are then an obvious problem since parasitic reflections from these surfaces always occur in such a way that instead of reproducing the characteristics of oscillation of bay and shelf, one has a tendency to excite the natural oscillations of the laboratory basin limited by the wall and paddle in which the model has been built.

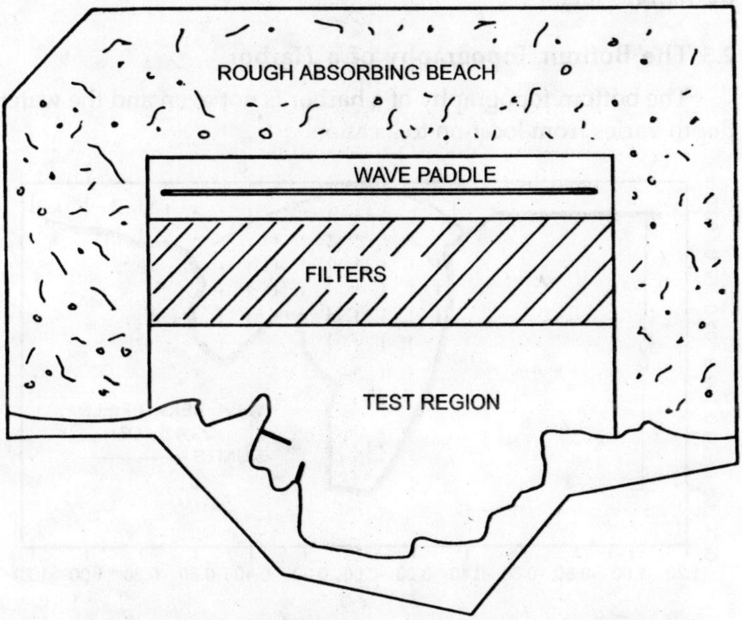

Fig. 2.2 : Scale model for studying long waves in a harbor would require reproducing the whole bay and shelf region, cumbersome wave absorbers and wave filters.

It is necessary to install wave absorbers or filters around the periphery so as to avoid resonance of the whole model. As an order of magnitude, the dimension of absorbers and filters must be like one wavelength to be effective so that the model space allotted to these items can be necessarily great.

In the case of long wave oscillation induced by surf beats or nonlinear interaction, the wave filters should be placed on the whole bottom of the model below the free surface. Short wind waves having a significant motion near the free surface only are seldom influenced by the filters and long wave components are damped as their motion is still significant in the deeper part of the model.

Finally, there are problems of adjustment of the wave characteristics. Wave direction, height and frequency must all be carefully controlled if the results are to be interpretable. Most sensitive is the adjustment of frequency since very small frequency change may cause large differences in the wave behavior inside the harbor.

2.3 The Bottom Topography of a Harbor

The bottom topography of a harbor is not even and the water depth varies from location to location.

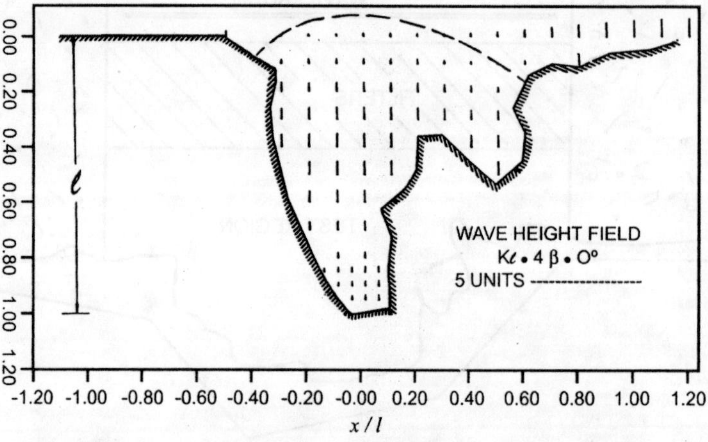

Fig. 2.3 : Amplification field in a bay of arbitrary shape assumed to be of constant depth. It is seen that the node cannot arbitrarily be located at the entrance.

The topography of a bay and its offshore area can be classified into three regimes: (a) shallow water region where variation of bottom geometry is large and usually has a significant effect on wave characteristics; (b) transitional region where bottom topography tends to be two–dimensional and the water depth is relatively deeper than in the shallow water region and small three-dimensional irregularities of the bottom topography can be neglected; and (c) the deep water region where bottom topography has little effect on the wave behavior. The geometry in the bay and near the bay is very complicated, so it is necessary to use numerical calculations to incorporate the bottom and shoreline effect variation. Such numerical calculations require a distribution of mesh points and must be sufficient enough so that the variations of bottom topography and the wave behaviour can be accurately represented. Limitation of the region of calculation must be small enough so that the calculation cost is feasible without affecting the accuracy of the results.

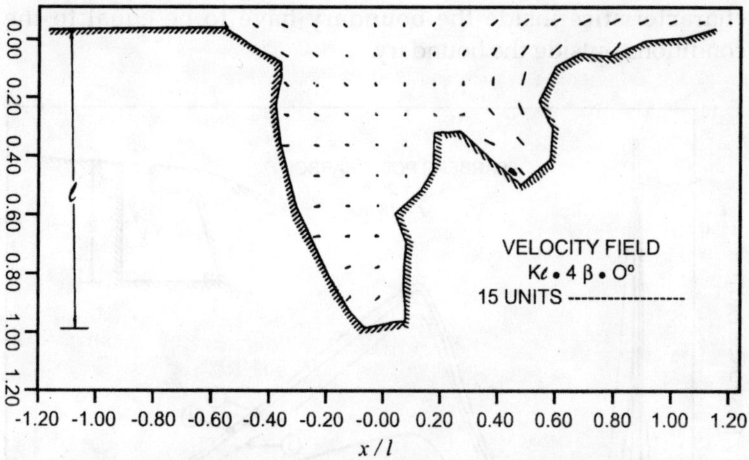

Fig. 2.4 : Instantaneous velocity field in a bay of arbitrary shape assumed to be of constant depth. It is seen that the smaller bay is subjected to resonant motion.

Since the behavior of waves inside the bay is related to incoming waves, the results obtained through the solution of the equation governing wave motion in the shallow water region without considering the nature of incoming waves is not sufficient. The outer region can be defined by an analytical solution obtained when considering the water depth as constant and equal to the depth at matching points. Then the analytical and numerical solutions are matched at the proper boundary. Using this scheme, incoming and reflected wave conditions will be properly considered.

Since the water depth used for the analytical solution is taken to be equal to the depth at the matching point, the amplification obtained inside the bay has to be modified by the effect of shoaling at the transition region; *i.e.*, the incident wave height at the matching location has to be multiplied by a shoaling coefficient. With the introduction given above, we now proceed to outline the theoretical developments.

In the outer region where the water depth is assumed to be constant, the governing equation is the same as presented in the previous section. The boundary conditions on the free surface, at the bottom and at infinity are the same except at the boundary where the matching occurs. At the matching boundary, the wave characteristics inside the boundary have to be equal to the conditions outside the boundary.

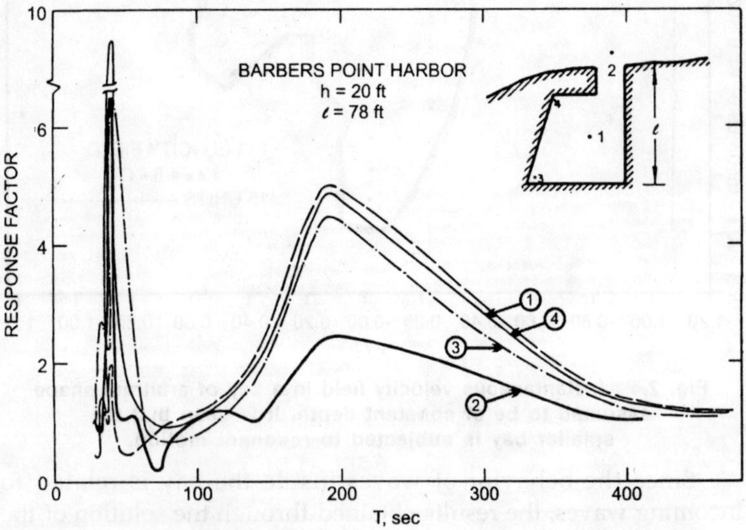

Fig. 2.5 : Response curves at four locations at Barbers Point Harbor. It is seen that the fundamental oscillation corresponds to the case where the basin is alternately filled and emptied with a practically horizontal free surface like a Helmholtz resonator in acoustic. Other peaks of the response curves correspond to initial oscillations.

2.4 The Oscilab System

The oscilab system is "bottom–oriented". The laboratory-habitat remains at its submerged location for the duration of the mission.

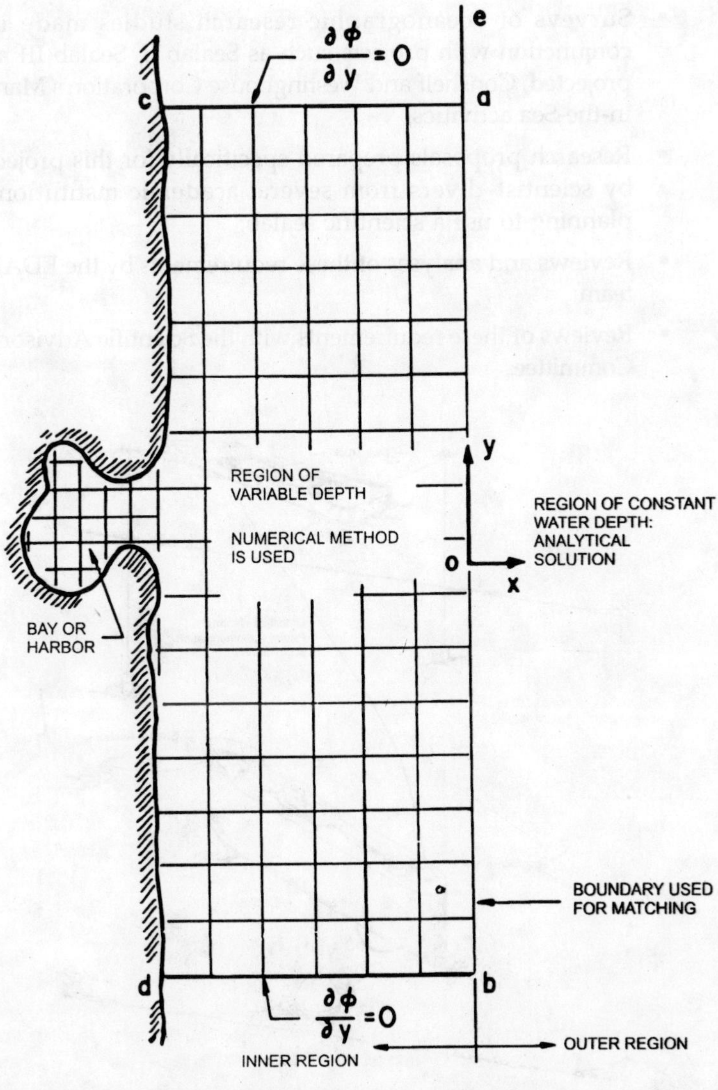

Fig. 2.6 : Schematic drawing is of a bay with reference coordinates.

The detailed conceptual designs of the Oscilab were based on two sets of considerations: the requirements of the scientist-diver and the designer's technical criteria.

Determining the requirements of the scientist–diver involved the following:

- Surveys of oceanographic research studies made in conjunction with projects such as Sealab II, Sealab III as projected, Conshelf and Westinghouse Corporation's Man-in-the-Sea activities;
- Research proposals prepared specifically for this project by scientist–divers from several academic institutions planning to use a scientific sealab;
- Reviews and analyses of these requirements by the EDAL team;
- Reviews of these requirements with the Scientific Advisory Committee.

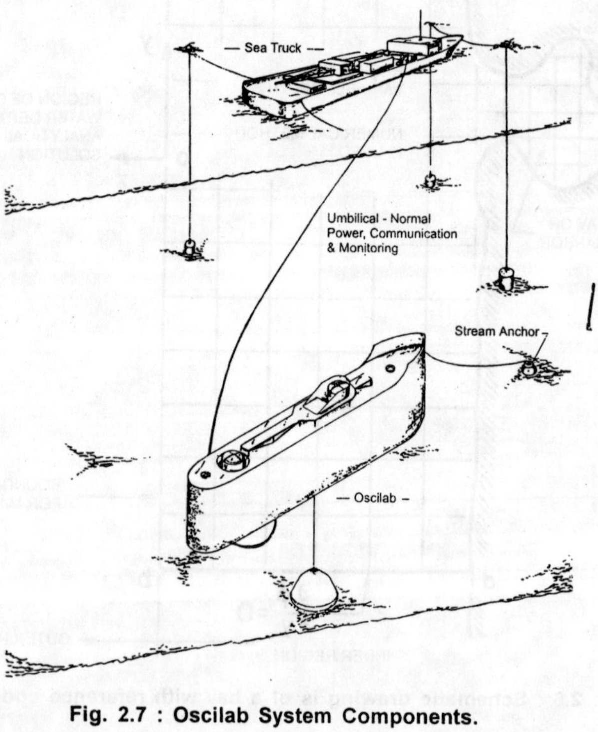

Fig. 2.7 : Oscilab System Components.

The oscilab systems include, the laboratory – habitat, the surface – support ship and the surface – subsurface linkages.

The following diagram indicates the interior arrangement of oscilab.

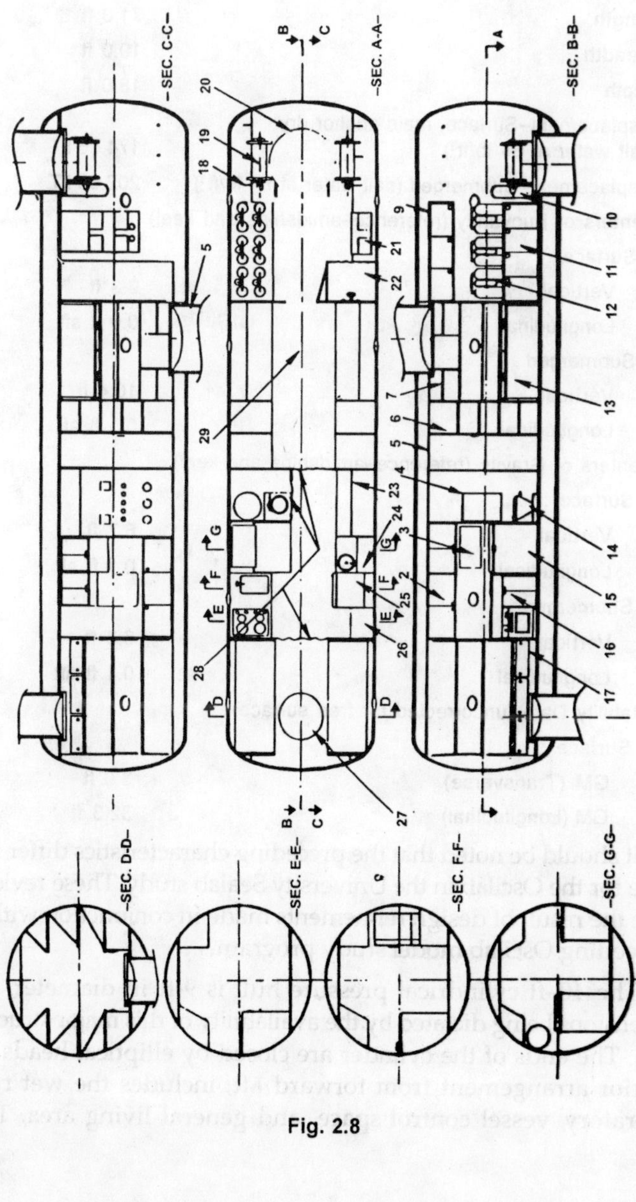

Fig. 2.8

The laboratory–habitat (the Oscilab) is a non-propelled vessel designed to permit six men to live and work continuously at a maximum depth of 300 ft for a two-week period under saturation-diving conditions. The Oscilab has the following principal geometric and hydrostatic characteristics:

Length	71.0 ft
Breadth	10.0 ft
Depth	18.0 ft
Displacement–Surface, main anchor dry (salt water at 64 lb/ft^3)	174.75 LT
Displacement–Submerged (salt water at 64 lb/ft^3)	202.65 LT
Centers of Buoyancy (reference–amidships and keel)	
Surface :	
Vertical	9.3 ft
Longitudinal	0.0 ft aft
Submerged :	
Vertical	10.4 ft
Longitudinal	0.1 ft aft
Centers of Gravity (reference–amidships and keel)	
Surface :	
Vertical	6.7 ft
Longitudinal	0.1 ft aft
Submerged :	
Vertical	6.3 ft
Longitudinal	0.1 ft aft
Stability Data (uncorrected for free surface)	
Surface :	
GM (Transverse)	3.0 ft
GM (Longitudinal)	32.3 ft

It should be noted that the preceding characteristics differ from those for the Oscilab in the University Sealab study. These revisions were the result of design refinements made in conjunction with the succeeding Oscilab model study program.

The 40–ft cylindrical pressure hull is 9-ft in diameter, such dimensions being dictated by the availability of this major structural item. The ends of the cylinder are closed by elliptical heads. The interior arrangement from forward aft, includes the wet room, laboratory, vessel control space, and general living area. Three access trunks penetrate the hull: a 48–in diameter wet room trunk; a 30–in diameter escape trunk in the living quarters; and a 48–in diameter trunk in the laboratory space which provides topside access and also serves as a lock. Other hull penetrations include those for viewports and electrical, mechanical, and piping leads. With the exception of the viewports and topside access trunk, all penetrations are located as near the bottom of the hull as possible to minimize the dangers of flooding and loss of breathing gas in the event of a hull–fitting failure.

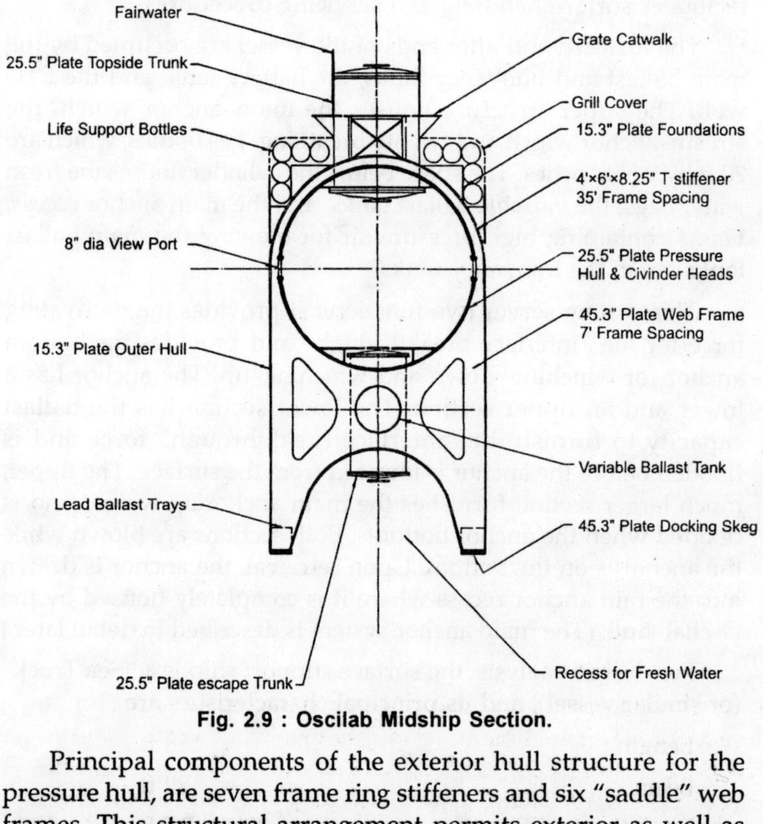

Fig. 2.9 : Oscilab Midship Section.

Principal components of the exterior hull structure for the pressure hull, are seven frame ring stiffeners and six "saddle" web frames. This structural arrangement permits exterior as well as interior loadings of the pressure hull and allows the use of a thin hull plating. Heavy longitudinal flat bars or keels are fitted at the bottom of the web frames, port and starboard, for dry–docking; and lead ballast trays are located on these flat bars.

The ends of the vessel are transversely framed. Twin vertical keels support the end ballast and buoyancy tanks, the battery, and the personnel transfer capsule (PTC) well structure.

An outer hull of light plating forms a fairing and protecting envelope for the pressure hull, external tankage, and equipment. The hull is "wall–sided" in way of the cylinder, but some "ship–shapeness" is provided at the ends to reduce towing resistance and enhance the Oscilab's sea–keeping characteristics. A small weather deck, protected by a bow spray–shield and access–trunk fairwater, facilitates surface handling and servicing procedures.

The forward and after ends of the vessel are occupied by the main ballast and buoyancy tanks, the battery tank, and the PTC well. The super structure houses the main–anchor winch, the stream–anchor winch and twenty breathing–gas bottles, which are 20 ft by 9 5/8 inches. The space below the cylinder houses the fresh water bags, the variable ballast tanks, and the main anchor recess. Flasks containing high pressure air for blowing the main ballast tanks are in the buoyancy tanks.

This system serves two functions: it provides the motivating force for the "interface breakthrough" and provides the bottom anchor for winching–down and winching–up. The anchor has a lower and an upper section. The lower section has the ballast capacity to furnish the "interface breakthrough" force and is flooded before the anchor is lowered from the surface. The upper, much larger section furnishes the main anchoring weight and is flooded when the anchor bottoms. Both sections are blown while the anchor is on the bottom. Upon retrieval, the anchor is drawn into the rain anchor recess where it is completely housed by the Oscilab hull. (The main anchor system is described in detail later.)

For design analysis, the surface support ship is a "Sea Truck" (or similar vessel), and its principal characteristics are :

Length	138 ft
Beam	29 ft
Draft (loaded)	about 7.5 ft
Clear deck area	102 × 29 ft

The ship's deck houses are confined to roughly the forward quarter length. The remaining three-quarters of the length is a flat working deck. This feature makes Sea Truck an extremely versatile support ship, since its deck load can be composed of a variety of modular arrangements–each arrangement meeting the requirements of a specific mission.

The design concept of Sea Truck also reduces port time and nonproductive port expenses to a minimum. Modular units (such as vans, decompression chambers and power units) can be prepared ashore before the ship's arrival. Then the units must only be hoisted aboard and secured to the deck. Units of heavy, concentrated deck loads may require local (and temporary) reinforcement of the ship's structure.

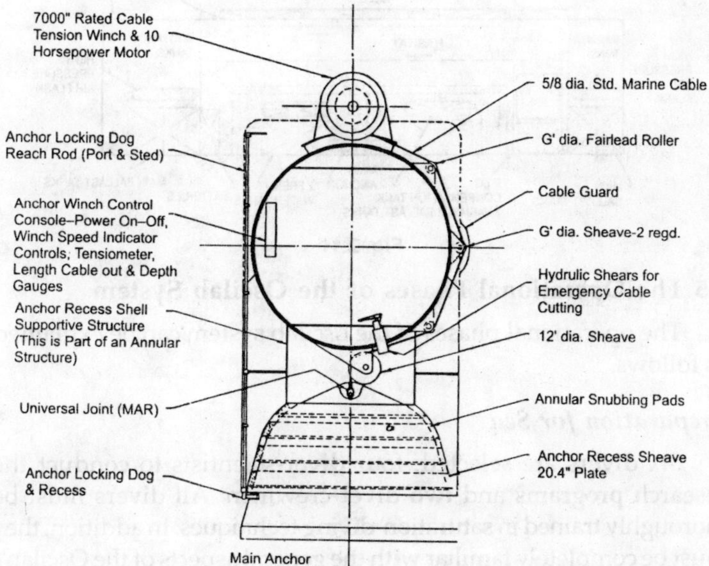

Fig. 2.10 : Oscilab Main Anchor System.

The Oscilab system requires that Sea Truck carry the following:

- Two Diesel generator sets, one for standby
- One reel and associated equipment for tending the umbilical
- One decompression chamber
- One deck crane for handling the PTC
- Communications and monitoring equipment (in a van)
- In-transit accommodations for divers and accommodations for any specialized surface-support personnel that may be required

- One 25- to 40-ft power work boat
- Miscellaneous equipment required to support the mission (in a van)

The following Diagram indicates oscilab propile and Deck plan.

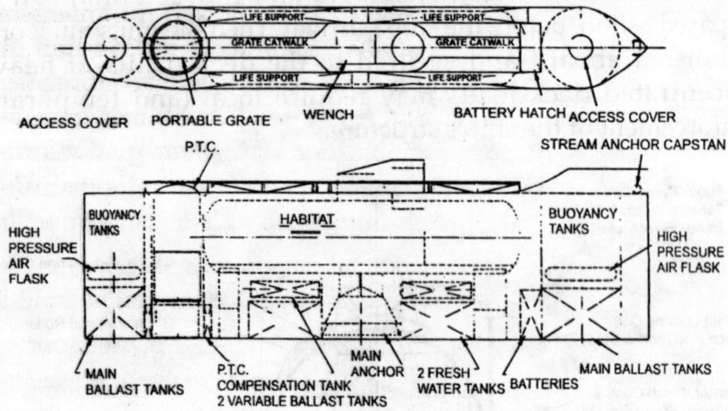

Fig. 2.11

2.5 The Operational Phases of the Oscilab System

The operational phases of the oscilab system can be explained as follows.

Preparation for Sea

Six divers are selected; four diver-scientists to conduct the research programs and two diver-crewmen. All divers must be thoroughly trained in saturation-diving techniques. In addition, they must be completely familiar with the general aspects of the Oscilab's operation and the details of emergency procedures. The two diver-crewmen must be thoroughly trained in all phases of the Oscilab's operation since this will be their primary duty. They should also be familiar with the broader aspects of the research to be accomplished so that they may effectively aid the diver-scientists. One of the diver-scientists is designated "chief scientist," and one of the diver-crewmen "captain." Regarding safety and conduct of the mission, the captain makes the final decision, but he is expected to be flexible. All major operational decisions should be cleared with the proper authority on the support ship if conditions permit.

The Oscilab is made ready for sea in all respects including the testing of all systems. To ensure safe and proper operation, the following procedure must be strictly observed in loading all items:

- All items must be stowed in their assigned locations.
- A record must be kept of the weights of all items.
- Weights must be kept within the design weight allowance.
- No item can be brought aboard which represents a potential hazard when exposed to a helium-oxygen atmosphere.
- When the Oscilab is fully loaded, the fore and aft draft must be carefully recorded by draft gauges and the weight and location of variable ballast water adjusted to place the Oscilab at its submerging trim waterline. The six divers are to be aboard and at their submerging stations for this procedure.

It should be emphasized that the variable ballast system is designed to permit flexibility in load weight and longitudinal distribution. A total loading condition outside the compensating capacity of the variable ballast system requires a weight and moment adjustment in the lead ballast. This adjustment can be made to a limited extent. However, the lead is located at the keel, and any lead removed to compensate for an added load weight will always decrease the stability. Conversely, a removal of load weight and an addition of lead will increase the stability.

The Oscilab is given an internal pressure test of 190 psi (using air) just before departure.

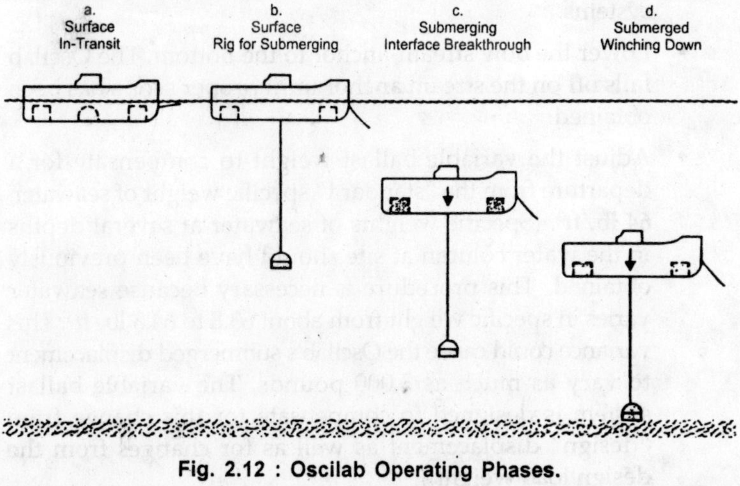

Fig. 2.12 : Oscilab Operating Phases.

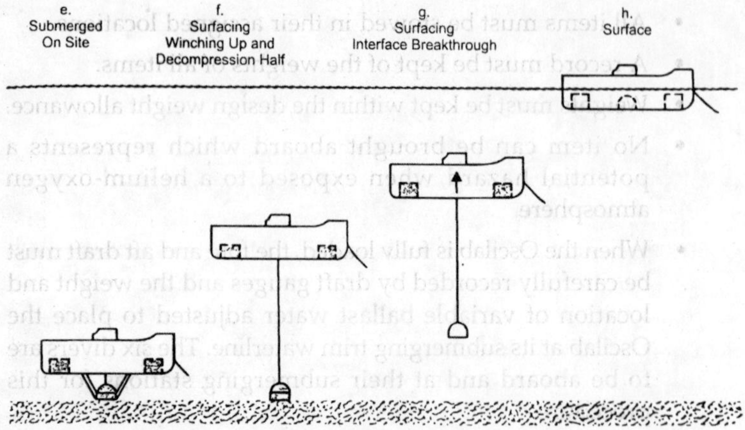

Fig. 2.13 : Oscilab Operating Phases.

2.6 The Preparation for Submerging

In preparation for submerging, the divers board the Oscilab and make a "dry" entry through the topside access trunk after the interior pressure has been reduced to 1 atmosphere. They then go to their submerging stations. The two diver-crewmen will be in the control space. Two diver-scientists will be aft in the living space and the other two diver-scientists in the wet room.

Oscilab. The "rig-for-submerging" procedure is as follows :

- Attach the umbilical and test all surface–subsurface systems.
- Lower the bow stream-anchor to the bottom. The Oscilab falls off on the stream anchor until proper scope has been obtained.
- Adjust the variable ballast weight to compensate for a departure from the "standard" specific weight of seawater, 64 lb/ft^3. (Specific weights of seawater at several depths in the water column at site should have been previously obtained. This procedure is necessary because seawater varies in specific weight from about 63.8 to 64.3 lb/ft^3. This variance could cause the Oscilab's submerged displacement to vary as much as 3,000 pounds. The variable ballast system is designed to compensate for this change from "design" displacement as well as for changes from the design load weight.)

- Cast loose the towing cable. This is done topside by a crew member of the support ship.
- Flood the lower part of the main anchor and lower it to within 40 ft of the ocean bottom.
- Close the topside trunk hatch, open the air-vent line, and start charging the Oscilab with helium. The Krasberg pO_2 meter will now feed the required amount of oxygen into the atmosphere. The Oscilab is pressurized to site-depth pressure.

Before the Oscilab's submergence, the support ship retrieves the towing cable and lays out a three or four-point moor on a course approximately parallel to the Oscilab's position and at a stand-off distance of 150 feet.

Submerging–Interface Breakthrough

"Interface" is the depth range 0 to 50 ft. This is the anticipated depth range for the severest surface wave action that could affect relative motion of the Oscilab with respect to the ocean bottom. In a seaway, a cable connecting a bottomed anchor to the Oscilab in the interface would be subject to excessive snap stresses and the possibility of kinking and breaking. In this design concept, the main anchor is lowered to within 50 ft of the bottom. Thus cable tension is maintained until the Oscilab is submerged to a depth of 50 ft and the anchor is bottomed. At this depth, there is very little relative motion between the anchor and the Oscilab; commensurately low dynamic stresses are placed on the cable.

All divers are at their submerging stations during the interface breakthrough. The diver-crewmen, stationed at the controls, vent the main ballast tanks. They anticipate the bottoming of the main anchor by activating the winch (at a slow speed) at a depth of 35 ft, thus minimizing cable slack due to inertia forces. When tensiometer readings show that the anchor is bottomed, the winch is stopped. Steady depth and tensiometer readings indicate that the system is in static equilibrium at 50 feet.

2.7 Submerged - On Site and Submerged–Emergency

When the Oscilab is on site, the divers check all surface-subsurface systems and scan the bottom and main anchor by television. When the pressure is equalized, the wet-room hatch is opened and one of the diver-crew makes the first excursion dive to examine the anchor and to connect the Oscilab to the anchor by preventer bars. The on-site routine can now begin.

On-site watch and duty bills are established. One member of the diver-crew remains aboard the Oscilab at all times for monitoring and to maintain communications with the surface ship and the divers working outside. The diver-scientists make excursions in pairs, accompanied by a diver-crewman if necessary.

A careful record must be kept of all on-site weight changes, such as changes incurred by bringing heavy specimens aboard and by the depletion of supplies. Adjustment in variable ballast is made to maintain approximately a zero load change. Load changes are checked by monitoring the cable tensiometer after the preventer bars have been removed just before ascent.

Oscilab. The longitudinal attitude (trim) of the Oscilab is independent of the ocean bottom topography. The main anchor may be inclined, but the Oscilab, tethered to the anchor and possessing a large amount of submerged stability, should have little or no trim. Preventer bars are secured between the Oscilab and anchor as cable back-up devices.

Support Ship. The support ship continues to supply normal power, monitor the Oscilab visually by television, monitor the Oscilab's breathing gas and maintains communications with the Oscilab.

There are five situations in which all divers leave the Oscilab immediately : if the support ship must leave position quickly because of weather and sea conditions, if there is hull leakage, if there is severe contamination of the habitat's atmosphere which cannot be rectified within the BIBB system time allowance, if the life-support system fails or if there is serious illness or injury.

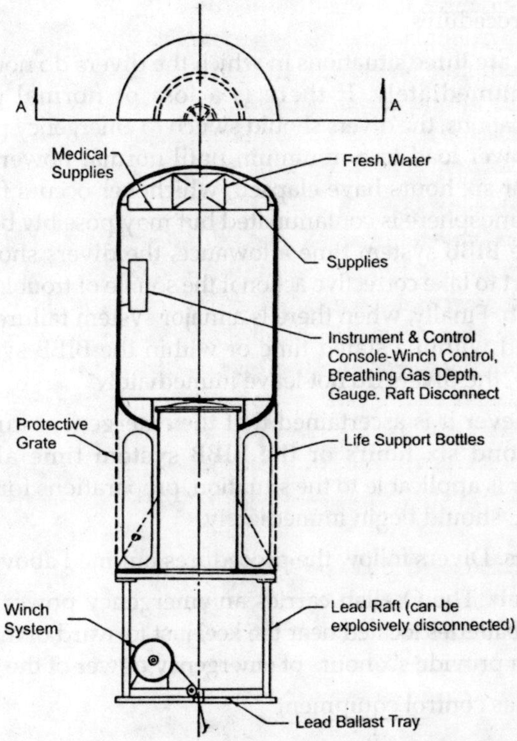

Fig. 2.14 : Personal Transfer Capsule (PTC).

The following actions should be taken for the preceding situations :

- Cut off all systems except the science refrigerator and secure the Oscilab to the extent that the situation permits. If conditions permit, switch the refrigerator to emergency power to preserve specimens until the Oscilab is reoccupied.
- All divers are to leave the Oscilab by the wet room or emergency hatches and make a wet entry into the PTC.
- The PTC is secured and its winch activated to pay out the cable for a controlled surfacing.
- The PTC surfaces well clear of the support ship and is towed by a work boat to the support ship. It is taken aboard by the support ship's crane and mated to the deck decompression chamber (DDC).

- Divers enter the DDC to complete decompression procedures.

There are three situations in which the divers do not leave the Oscilab immediately. If there is a loss of normal power or communications, the divers should switch to emergency power and reduce power load to a minimum until normal power has been restored or six hours have elapsed, whichever occurs first. If the Oscilab atmosphere is contaminated but may possibly be rectified within the BIBB system time allowance, the divers should make every effort to take corrective action if the source of trouble is aboard the Oscilab. Finally, when there is a major system failure that may be rectified within a stated time or within the BIBB system time allowance, the divers do not leave immediately.

Whenever it is ascertained that the emergency situation will exist beyond six hours or the BIBB system time allowance, whichever is applicable to the situation, preparations for surfacing in the PTC should begin immediately.

Divers. Divers follow the procedures outlined above.

Oscilab. The Oscilab carries an emergency power source of lead-acid batteries located near the keel just forward of the pressure hull. It can provide six hours of emergency power of the following:

1. Gas control equipment
2. Sonic transmitter
3. 500 watts for emergency lighting
4. Heated suits
5. Refrigeration

If items (4) and (5) above are eliminated, the emergency power would last 20 hours.

The Oscilab can be entered for emergencies or other reasons through the topside or wet-room locks. This can be done very easily by a person using shallow diving gear when the Oscilab is near the surface.

Support Ship. The support ship's principal emergency activity is to retrieve the PTC, hoist it aboard, and mate it to the DDC. The Oscilab should be immediately notified of any potential or existing factor on the surface that might contribute directly or indirectly to an emergency situation.

2.8 The Principal Elements of the Oscilab Design

The principal elements considered in the design of the Oscilab are

- Hull design
- Main anchor system
- Communications, power and lighting
- Life support systems
- Diver mobile equipment.

Hull Design

The following discussion of the Oscilab hull design includes structure, materials, buoyancy, stability, a condition summary table and interior general arrangement.

Structure

This report discusses only the principal structural item, the pressure hull. The scantlings and other structural items are estimated for weight, buoyancy, and stability studies.

The pressure hull design involves two major considerations, geometry and strength. The geometry is concerned with size and shape. In the interest of low breathing-gas costs, the size is just large enough to provide for adequate working and living spaces. The shape is a function of interior gneral arrangement as well as of hydrodynamic considerations. A cylinder was chosen for this design because great depth and associated strength requirements are not present.

Strength requirements depend largely upon the interior and ambient sea pressure differentials and the "directions" of these differentials with respect to their manner of loading the pressure hull. A summary of pressure differentials, Δp, and their directions are as follows:

- Depth-0 to on-site depth (300 feet is maximum); Oscilab pressurized to site-depth on surface.

  Δp–133 psi

  Direction-internal (material in tension)
- Depth-on-site depth

Δp-0 at interface to about 6 psi at top of hull

Direction–internal (material in tension)

- Depth–on-site to 40 ft (start decompression halt)

  $\Delta p_{(max)}$–115psi

  Direction–internal (mterial in tension)

- Depth–40 ft (end of decompression halt)

  Δp–18psi

  Direction–external (material in compression)

- Depth–0 ft (if decompression is on surface)

  $\Delta p_{(max)}$–133psi

  Direction–internal (material in tension)

The greatest external Δp is 18 psi, and the greatest internal Δp is 133 psi. These two values must be compared with critical values derived from presssure-hull design data to check the structural adequacy of this data.

Main Anchor

The main anchor, has the shape of a truncated cone capped by an elliptical dome. This shape was chosen mainly because the main anchor must be drawn by a single cable into the recess at the bottom of the Oscilab. Thus, it must be circular in corss section and provided with sloping sides to ensure that it will enter the recess without fouling and to provide for snugging against suitable pads so that it will ride in a housed condition without bumping. This shape also increases the stability of the bottomed anchor.

Anchor Ballasting. The anchor is divided into upper and lower sections, each containing four ballast tanks. The lower section is completely flooded while the Oscilab is on the surface. It provides the net downward force for the interface-breakthrough phase of descent.

The main-anchor operations can be explained as follows.

Surface-in transit. The dry anchor is housed in the anchor recess. It is held in place by the cable and by dogs connected by reach rods to topside hand-operated levers. A short electric cable connects the Ocilab with the lower section solenoid vent valves.

Surface-rig for submerging. The lower section of the anchor is flooded by energizing the vent-valve solenoids. The anchor is then lowered to 50 ft above the ocean bottom. The short electric cable pulls free of the Oscilab in the first few feet of the anchor's descent.

Submerging–interface breakthrough. When the main ballast tanks are flooded, the main anchor, with the lower section flooded, sinks to bottom, pulling the Oscilab through the shallow interface depths. An electro-mechanical or pressure sensitive linkage is tripped at bottom contact, opening the upper section vents and allowing the upper section to flood.

Submerged–on site. The main anchor is fitted with pad-eyes for securing preventer bars between the anchor and the Oscilab. These bars, attached by the first diver out, preserve fore and aft trim and serve as cable back-up devices.

Submerged–rig for surfacing. The preventer bars are removed and all vents on the main anchor are closed manually. Two 400-ft light electric cables are connected to the Oscilab and the main anchor with the slack flaked out on the bottom. One cable serves as a back-up and both are capable of energizing all solenoid-operated blow valves and starting the water pump.

Surfacing-interface breakthrough. The solenoid-operated blow valves are energized, and the upper and lower section ballast tanks are blown with high pressure air stored in air flasks in the anchor. The blown ballast issues water through nozzles designed to remove mud if the anchor has lodged in a soft bottom. The nozzles are provided with special float valves that will pass water but retain air. All possible low and high points in each tank are vented separately so that the tanks will drain and vent completely and any degree of tilt will be short of overturning.

If the anchor is not freed when the ballast is blown, a high-pressure water pump can be activated. This pump operates high-pressure water jets on the underside of the anchor to help free it. In addition, a "bubble" of air can be injected into the main ballast tanks to loosen the anchor from the bottom. Immediately after the anchor begins to ascend, these tanks should be vented so that ascent will not be too fast. As son as the anchor is lifted clear of bottom, the two light electric cables are cut loose.

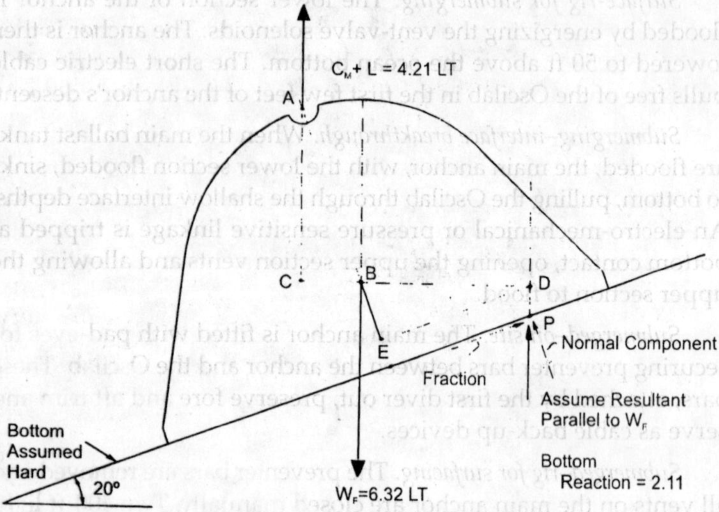

Fig. 2.15 : Oscilab Anchor Stability.

Surface. If sea conditions permit, scuba divers inspect the anchor and cable for fouling when the anchor is within a few feet of Oscilab. The electric cables are unplugged and retrieved on the Oscilab topside deck. In adverse seas, no attempt is made to salvage the electric cables. Since the anchor and cable rotate, these electric cables have little opportunity to become entwined around the main cable when falling away from the Oscilab as the anchor is lifted off the bottom. Ideally, these cables will fall clear of the anchor as it rises. Some of the cable may be looped across the anchor; however, this possibility is diminished by the bee-hive shape of the anchor. Nevertheless, it would not be a serious handicap to the anchor housing since the snubbing pads have some resiliency and can accommodate the light cable to a certain degree.

When the anchor is housed, the high-pressure line is brought topside by scuba divers where it is uncapped and connected to a source of high-pressure air. This is done to avoid getting salt water into the airlines, which might occur if an under-water connection were attempted. The anchor battery pack is replaced.

Now let us study about the winch.

The following considerations govern the choice of the winch, the winch drive and their locations.

Lifting capacity. The greatest demand on the winch system lifting capacity occurs during the winching-down phase of submerging. This is a dynamic condition in which the cable tension, C_D, is 1.88 long tons or 4,210 pounds. The winch chosen, with a rated lifting capacity of 7,000 lbs., provides a margin of 2,800 lbs. or 40%. The drum can hold 400 ft of 3/4-in. marine cable. The winch system (excluding the cable) weights about 2,500 lbs; requires a space 6 ft long and 3 ft in diameter; and costs $5,000.

Powering. Two possible sources of winch power are compressed ari and electricity. Compressed air drivers are very inefficient and therefore demand large quantities of compressed air, which is a luxury in this design since all high-pressure air is carried aboard the Oscilab. On the other hand, electric power is supplied from the support ship under normal conditions, and electric motor drives are more efficient than air drives. A 10-horsepower electric motor which operates under ambient pressures in an oil bath was therefore selected.

Location. The winch and winch drive are located topside just aft of the topside trunk to facilitate maintenance and cable inspection while the Oscilab is on the surface.

The Oscilab communications system consists of links between the Oscilab and the support ship and between the Oscilab and the diver.

There are audio and visual links between the Oscilab and the support ship, including interior and exterior environment sensing and recorder displays, and a back-up acoustic communication unit. An audio communications link is provided between the Oscilab and the diver; the diver's hookah gear umbilical contains the cable.

Communication and instrumentation equipment required are listed below.

- Oscilab

a. Intercom and speech unscrambler-two-way intercom with an integrated speech unscrambler linked with the support vessel and the diver's communication unit

- Electrowriter–means of transmitting diagrams and figures
- TV camera–camera for observing the Oscilab interior from the support ship

- TV camera and monitor–equipment for observing and monitoring the site environment
- Oscilab interior environment instrumentation–chart-recording equipment for wet and dry bulb temperatures.
- Acoustic communications unit and transducer–equipment for emergency communications between the Oscilab and support ship
- Environment sensor system–equipment that provides continuous chart-recording of site environmental characteristics such as temperature, pressure, current speed and direction, ambient light, pH and turbidity

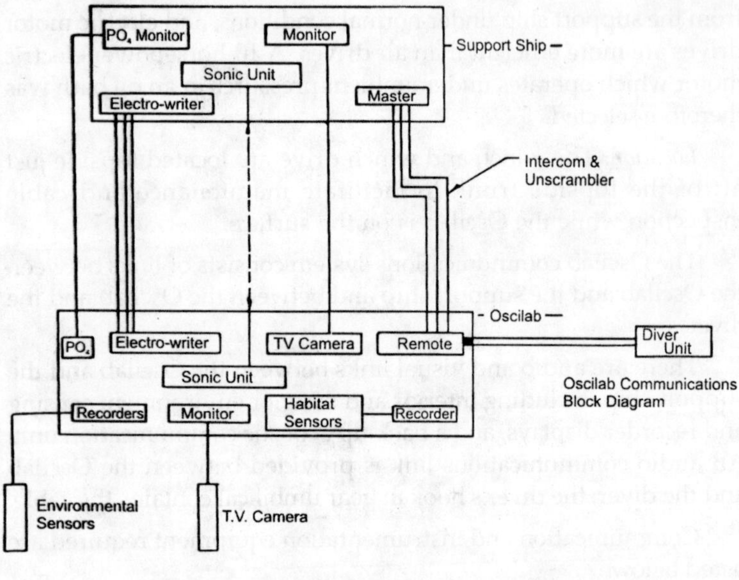

Fig. 2.16 : Oscilab Communications System.

Now let us learn about Power and Lighting and also about the Life support systems.

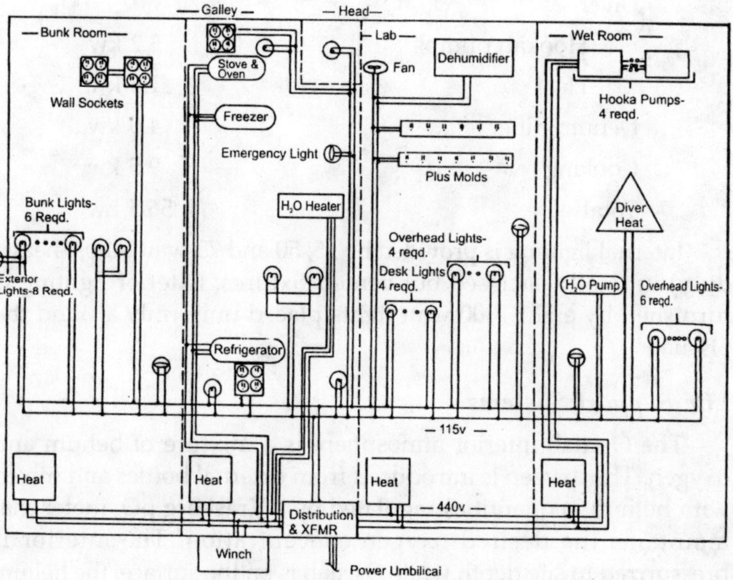

Fig. 2.17 : Oscilab Electrical System.

The Oscilab system has two sources of electrical power, normal and emergency. The normal source of power is a 60-kw., 3-phase, 450-volt generator located on the deck of the support ship. The power is delivered to the Oscilab through cables in the umbilical. A second generator serves as a standby unit. The emergency source of power is carried on the Oscilab and consists of a 150-kw-hr lead–acid battery pack.

Its power requirements can be divided into the following categories:

Lighting	9.6 kw
Communications	0.5 kw
Heat (interior)	13.0 kw
Heat (hot water)	4.5 kw
Refrigerators-freezers	0.2 kw
Fans	0.1 kw
Pumps	0.4 kw
Winches	10.0 kw

Diver	
Hookah pumps	3.2 kw
Heat	1.5 kw
Dehumidifier	4.0 kw
Cooking	9.5 kw
Total	56.5 kw

Internal lighting is provided by 25, 50 and 75-watt bulbs (heavy duty) in totally enclosed but vented fixtures. Exterior lighting is furnished by eight 1000-watt lights placed uniformly around the Oscilab.

Life Support Systems

The Oscilab interior atmosphere is a mixture of helium and oxygen. The oxygen is introduced from external bottles and mixed with helium in quantities regulated by a Krasberg pO_2 meter that maintains the desired oxygen concentration. The interior is pressurized to site depth while Oscilab is on the surface; the helium is supplied by the support ship. When the Oscilab is on site and the lower trunks are open to the sea, additional helium is introduced from external bottles to compensate for gas losses and to maintain a constant water level in the trunks. A float control in the trunk regulates the introduction of this makeup helium. The atmosphere is circulated through a purification system to remove water, carbon dioxide, hydrocarbons and other trace contaminants.

In the design of the mixed-gas system, the gas-loss-versus-reclamation ratio during the decompression process was considered. The cost of the gas required to pressurize the chamber is under $1,160. The cost of a gas recovery and purification system, estimated to be about $40,000, precluded this alternative. Hence, the gas will be discharged directly into the air if decompression is carried out at the surface and into the sea if this procedure is accomplished at depths of 40 to 50 feet. In the latter case, a pump or surface-buoyed gas line must be used during the final stages of decompression.

The divers use a closed hookah system which delivers breathing gas from the Oscilab's interior and returns it for purification. In addition, each diver carries emergency bottles of breathing gas. A BIBB system is installed for emergencies occurring within the Oscilab.

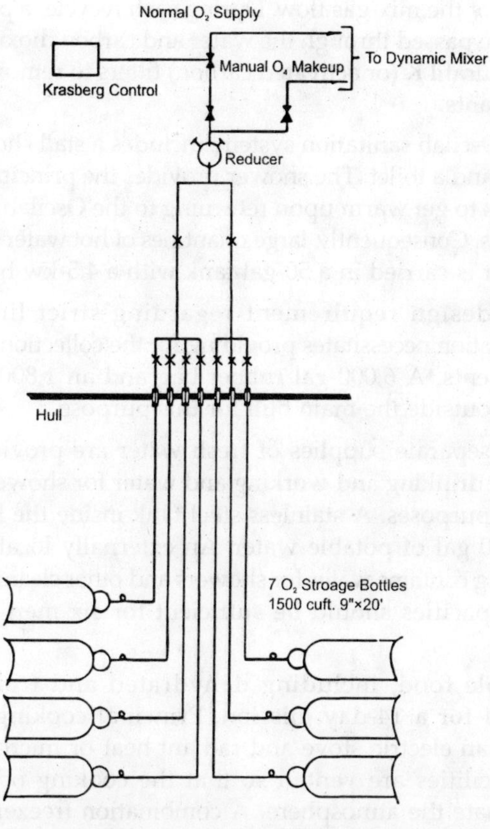

Fig. 2.18 : Oscilab Oxygen System.

Now let us try to learn few more things about Heating and Ventilation system of Oscilab.

The Oscilab interior heating system is designed so that there are no cold surfaces to absorb radiant energy from the occupants. Heat is provided by imbedded cables in the floor and radiant panels on the sides and overhead. Three kilowatts are provided for floor heating, 7 for side heating, and 3 for overhead heating. The floor elements would be thermostatically controlled as would be the individual radiant panels of the side and overhead system.

The Oscilab interior ventilation system, consists of ducts with intakes near the floor and several outlets near the overhead. The occupants can adjust these outlets to vary the quantity and direction of the mix-gas flow. During each recycle, a portion of the gas will be passed through the water and carbon dioxide removers and the Purafil K (or activated carbon) filters to remove odors and trace irritants.

The Oscilab sanitation system, includes a stall shower, a small lavatory, and a toilet. The shower provides the principal means for the divers to get warm upon returning to the Oscilab from on-site excursions. Consequently, large quantities of hot water are required; this water is carried in a 50-gal tank with a 4.5-kw heater.

The design requirement regarding strict limits on site contamination necessitates providing for the collection of all wastes and effluents. A 6,000-gal rubber bag and an 1,800-gal bag are installed outside the main hull for this purpose.

Two separate supplies of fresh water are provided: potable water for drinking and working and water for showers and other cleaning purposes. A stainless steel tank inside the hull contains about 170 gal of potable water. An externally located 6,000-gal rubber bag contains water for showers and other cleaning purposes. These capacities should be sufficient for six men on a 14-day mission.

Ample food, including dehydrated and frozen food, is provided for a 14-day mission. Principal cooking equipment includes an electric stove and radiant heat or micro-wave oven. These facilities are vented so that the cooking odors will not contaminate the atmosphere. A combination freezer-refrigerator provides for food preservation.

The PTC has a completely independent life-support system capable of supporting six men for 6 hours under emergency conditions. Breathing gas is contained in two size A helium bottles and one size A oxygen bottle. A blower and duct system will circulate the gas through baralyme and Purafil K for carbon dioxide and hydrocarbon removal. No provision is made for moisture removed or heating. Emergency kits of water, food and medical supplies are stowed in the PTC.

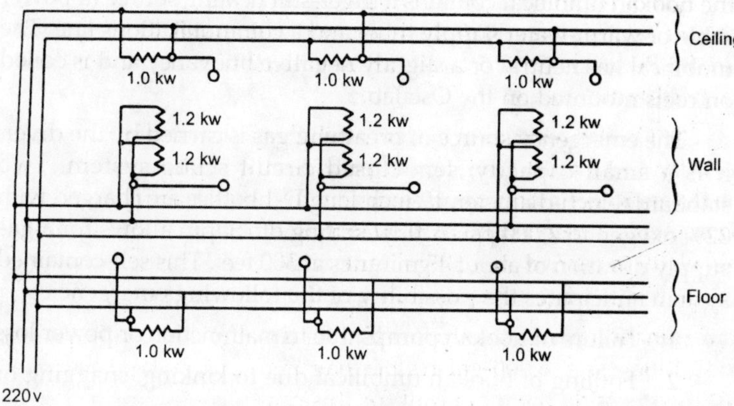

Fig. 2.19 : Oscilab Heating System.

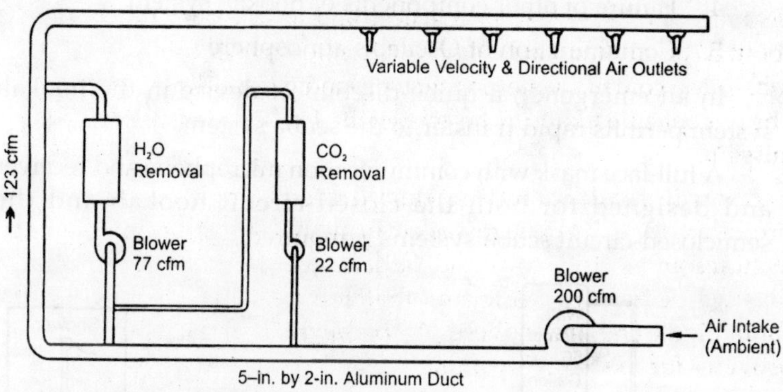

Fig. 2.20 : Oscilab Ventilating System.

Finally, it is needed to learn about diver Mobile Equipment. Let us start with breathing Apparatus.

The diver's breating gas may be derived from two sources while he is on site excursions away from the Oscilab, the normal source and the emergency source.

The mormal source of supply uses a closed-circuit hookah system in which the gas is pumped from the interior through one line of a double hose to the diver. After inhalation, the diver exhales into the other line of the hose, and the gas returns to the Oscilab for purification and revitalization. In addition to the double hose, the hookah umbilical contains a diver-suit heating source (a power cable or warm-water supply tube) and a communications line. The umbilical has neutral or a slightly negative buoyancy and is coiled on reels mounted on the Oscilab.

The emergency source of breathing gas is carried by the diver; it is a small capacity, semiclosed-circuit scuba system. Two standard 6-inch-diameter, 19-inch-long D-1 bottles are charged with 22% oxygen to 2,000 psi. A flow setting of 35 lpm allows for a gas supply duration of about 45 minutes at 300 feet. This self-contained system anticipates the possibility of the following emergencies:

1. Failure of hookah pumps due to malfunction or power loss
2. Fouling of hookah umbilical due to kinking, snagging or collapse
3. Cutting of the umbilical
4. Failure of other components of hookah system
5. Contamination of Oscilab's atmosphere

In an emergency, a quick-disconnect device in the hookah system permits rapid transfer to the scuba system.

A full-face mask with communication microphone and receiver and designed for both the closed-circuit hookah and the semiclosed-circuit scuba systems is required.

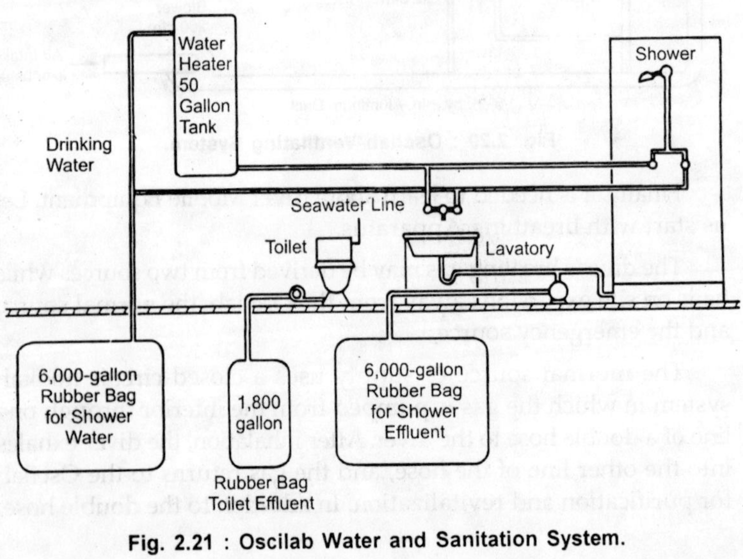

Fig. 2.21 : Oscilab Water and Sanitation System.

Two types of diver suits, the electrically heated and the warm-water heated suits were considered. Despite the relatively high cost of electric suits, it did not seem feasible at this time to use warm-water suits because this system requires 30 to 50 kws of energy for two suits. Electric suits require 1 to 2 kw of energy for two suits. A diver-mounted, electrically-powered, compact heat exchanger and pump could make the warm-water heated suit more attractive, but such a system is not practically available. Consequently, an electric resistance-wire heated suit with power supplied from Oscilab at 24 volts (AC) is used.

Another advantage of the electric suit is that this system eliminates the discharge of heated water into the site environment. In addition, an umbilical containing an electric cable is less bulky than one containing a warm-water supply tube or hose.

Additional diver equipment including fins, watches, compasses, and emergency sonic homing devices are similar to conventional types. The equipment was designed to operate in a high-pressure helium atmosphere.

Chapter Three
Underwater Human Activity

3.1 Introduction

The most exciting aspect in considering any underwater activity is that man may be able to accomplish the activity with ease, with pleasure with comfort and with profit. How well man be able to do this will depend on his ability to take advantage of the current information and technology.

Commercial and scientific underwater human activities have always been limited by the restrictions imposed by the underwater environment. Either the visibility is poor, the current strong, the temperature low, or the man is faced with a time limitation on his useful exposure in the water. A prolonged period of decompression may be necessary to leave the work site after an operationally useful period of time. Almost all scientific and commercial diving is done by what is now called the *bounce dive* technique. With this technique the worker, regardless of his purpose, enters the water, descends to the work site, spends a finite period of time there, and then returns to the surface according to a schedule determined by the depth and duration of his exposure at the work site.

If for the moment we ignore the limitations of visibility, temperature, current etc., and restrict ourselves to the problems of the depth of the work site and the duration of work on location, the working ability of each individual is severely limited, because he either stays only a short time on the bottom to avoid prolonged decompression, or he pays the price of an extensive decompression schedule.

The sea volume is estimated at 330 million cubic miles. Fresh water and estuarial volumes may be added to this huge space available to man for future underwater pursuits. Ninety-eight per cent of the sea is no deeper than 20,000 feet, bringing it within present technological skills.

In comparison, the air around the earth, which is easily and economically accessible to man, has a volume of 6,573 million cubic miles when extended to 10 miles altitude. At four miles, the volume is still vast – 5,590 million cubic miles. The volume of the air is 17 times that of the oceans. That difference is not a gauge of importance.

The oceans contain food and minerals and provide recreational opportunities. The oceans cover petroleum and mineral deposits that do not occur in the atmosphere surrounding the earth. Next to trains, water transportation is the best we have for economic bulk movements. These advantages of water space assure man's future underwater. It's all there, the opportunity exists, the technology will soon be developed.

Man has a future underwater either as an occupant of water or as an occupant of a vehicle or habitat contained within the water. This permits him the possibility of entering the water and pursuing his activities at sea level pressure in a specifically designed vehicle, or at ambient pressure related uniquely to water depth that he wishes to explore. If he is to use the vehicle whose own cabin atmosphere is maintained at one atmosphere, or sea level pressure, his health and physiologic future will be determined by the ability to maintain the artificial gas environment within the vehicle. We currently have rather considerable experience in doing this, obtained from the successful Fleet Ballistic Missile Submarine (FBM) Program.

Given the proper engineering resources and relatively unlimited power, we can maintain an artificial environment that appears to be harmless to man for extended periods of time. The odds are that it will not be the abnormal or artificial environment at only sea level pressure that limits man's exposure in the confined and closed ecologic system.

Now let us try to understand about the problems of sea level exposure of man in a closed habitat. The more interesting problem will be man's operations in the environment uniquely related to the ambient pressure of the water at the depth that he desires to attain. Current diving techniques are limited economically by decompression times when deep and/or prolonged exposures are conceived. Scientific activity under these limitations is tremendously expensive and far beyond the financial resources of any of our universities or non-profit organizations, even though these are principally financed by the Federal Government.

This negative approach is used only to introduce a note of caution and to justify an extension of opportunities that do exist that will require only modest funding to permit considerable activity. At the present time, within the limits of our engineering skills, there are many man years of effort possible in shallow water. These will permit us to anticipate virtually unlimited activity as our engineering skills advance. The safety of the operation remains the prime concern of the task analyst.

The ideal situation will be for man to enter whatever hyperbaric environment he wishes for as long as he wishes, and to leave this environment at his own convenience without interrupting his actual endeavors. For the scientist this appears to be quite possible. The scientist, if he is a marine biologist, wishes to be in the underwater space to directly observe the animal or plant life under study. He also wishes and needs the opportunity to select those specimens present at various times of the day in specific locations and the rapid opportunity of improving and correcting his collections. This can only be easily accomplished if he is able to stay at the site of sample collection.

3.2 The Saturation Techniques

At any particular depth, the amount of gas that enters the tissues is determined by the solubility of the gas in the various tissue compartments, the rate of diffusion into the compartments, the perfusion or delivery of the gas by the blood stream to the compartments, and the length of time exposed at the depth. Unfortunately, all tissues do not have equal perfusions, diffusions, or solubilities and therefore become saturated at different rates. The variability in tissue saturation rates can be described as tissue *half-times,* an expression found in every modern technique used to calculate decompression schedules.

Fortunately, the slowest possible tissues, those that take the longest time to become completely saturated at a particular pressure, probably do not have half-times in excess of six hours. The number of more rapid tissues, with respect to their half-times, is infinite, but for purposes of convenience, a limited number are usually included in the decompression schedule calculations.

If it is true that there are no tissues with significant half-times in excess of six hours, then one can assume that at any depth of exposure after 12 hours all tissues in the body are completely saturated and that all additional exposure time at that depth cannot influence the decompression schedule. Under these circumstances it would then be possible to stay for an unlimited time, at the depth chosen without significantly modifying the decompression schedule imposed by saturating the body tissues in the first place.

At the present time we can make dives without imposing decompression upon ourselves. A schedule of these dives is called the *no-decompression schedule* and varies at the present time by US Navy standards from several hours at 30 feet to 5 minutes at 150 feet. An implicit assumption made in all the no-decompression schedules is that the dive is begun with all body tissues saturated at one atmosphere, and that within the duration of the dive exposure to the depth elected, no decompression burden is imposed. Obviously, if either the depth or duration is exceeded, decompression would be required.

These same assumptions can be made, regardless of the starting saturation of the tissues, whereby no-decompression excursion dives can be made from any point of saturation previously decided. For example, in the initial plans for *Sealab III*, the habitat was scheduled to be at 450 feet where total saturation would occur. Excursion dives with little or no decompression were planned to 600 or 650 feet. Recent plans for *Sealab III* will locate the habitat at 610 feet with excursions to 850 feet.

The US Navy, at the Experimental Diving Unit, has just recently completed a six-day chamber dive under saturation conditions at 600 feet with five men. The following week, a second group descended in the chamber complex to 600 feet. Then two men made excursions to 825 feet for specific performance tests. The entire complex was then, moved to 825 feet simulated depth. Two days later, two men made 13-minute excursions to 1,025 feet without any apparent difficulty. Decompression was started the following day and was completed uneventfully. It was reported in the press that the problems of humidity in the dry chamber were eliminated with conditions maintained at a steady 85 degrees F and 60% relative humidity. In a recent test by International Underwater Contractors, Inc., in cooperation with AIRCO at their Central Research

Laboratory in Murray Hill, New Jersey, a saturation exposure to 600 feet was made by two men who then did excursion dives to 800 feet for two hours. This imposed on them a decompression schedule before they could return to the 600-foot level as they wished. One of the men, however, did make a five-minute bounce dive without imposed decompression to 900 feet from the 800-foot level. (During the week of March 11-18, 1968, the same research team went to 1,000 feet for 120 minutes.)

There is a tremendous advantage when performing protracted work in deep water if one can live in the Hyperbaric environment or live near the diving work site. Economically this can be a tremendous advantage because the logistics of diver support can be reduced and the considerable time spent in decompression schedules for each brief exposure at great depth can be saved until all the work is completed by all divers. The entire work schedule need not be interrupted hour by hour and day by day. The risk of having to decompress a group of men simultaneously, but only once, appears to be considerably less than having to decompress each man 20 to 30 times over a 40-day work schedule.

Some of the obvious advantages of saturation exposure compared to the usual bounce dive technique for underwater work can be listed as follows:

- A longer time is permitted at the work site.
- Less time is spent in decompression per total unit work time.
- There are fewer pressure changes per unit of work; therefore, considerable safety is added with respect to the possibility of decompression sickness.
- Excursion diving is possible with or without decompression, greatly increasing the range of useful depth available to the worker who starts from the saturation point.
- The saturated underwater worker has a considerable mobility that can be used for exploration.

3.3 Underwater Activity

The membership of the Underwater Society of America, which is made up of many noncommercial and nonprofessional divers yields an overlapping figure of possible involvement in purely recreational activities. This group of enthusiasts need a number of things for which we will ultimately have to assure provision. They need good supplies of reliably pure compressed air, they need relative security from surface hazard in the diving area, and ideally they need areas for recreational diving that will not be hampered by or inhibited by commercial activities.

Two underwater parks have already been created, one off St. John in the Virgin Islands, another off Key Largo in Florida, and one off the Coast of California is planned entirely within the confines of territorial United States. Many others will be required before long, not only in the United States, but throughout the world because the number of individuals who wish to skin dive is increasing rapidly. Water-surface recreation will require even greater long-range planning since many more million people are involved, and available resources are already restricted.

In scientific activities underwater, immediate advantages become obvious to anyone who has attempted any type of field work. If you wish to study a problem at 400 feet and you happen to be living there, when you want a new specimen, or you wish to go look again, or make another current measurement, or measure the visibility, or obtain a sample for chemical analysis, you just pop out of the habitat, obtain your sample, and go back into your laboratory. If you are working at the surface, it would be a 12- or 15-hour exercise by the time you got everything arranged, got into your diving suit, made your descent, obtained your sample after you have found where you were, if at the right place, and then went through decompression.

An advantage may occur whenever the quality or the importance of work to be done underwater reaches a critical point. Dr. J.B. MacInnis of Ocean Systems, Incorporated, discussed this critical point at the San Diego Conference of the Marine Technology Society. I will touch only on a few features that deserve serious consideration. One of these is related to the cost of men per hour. For example, if 100 hours of underwater work are required and the work site is such that the decompression schedule for 30 minutes bottom time is eight hours, then 100 hours of bottom work will require 1,600 hours of decompression, if each diver decompresses alone. If you are able to double up on the decompression of the men, you could reduce this to 800 hours.

However, since we do not have techniques for placing 200 people on the bottom at the same time so that each can perform 30 minutes of work and then safely decompress, a more reasonable approach is with saturation methods. We can have four or five people working on the bottom at a time, keep them saturated, switch work teams, and then decompress the lot in a large chamber on the deck of the work barge.

The Westinghouse Cachalot System uses a two-man submersible decompression chamber (SDC) for transferring the divers from the work site to a deck decompression chamber where they live during the work period and then safely decompress one time when the work is completed. As of June, 1967, they had used this saturation system on three separate jobs for 5,000 hours of productive work under pressure in depths beyond 200 feet. The work schedules were such that completion could not have occurred without the availability of the saturation technique.

The costs are high with all saturation systems because of the set-up time, gas expenses, and nondiving personnel required for continued safe operations. Current estimates, usually based on guaranteed minimum underwater working times of about 150 hours per week, are in the neighborhood of $150,000 per month for working depths around 200 feet. Greater depths would increase risk and gas expense since completely closed gas systems are not available.

The second economic feature deserving consideration is the cost of alternative methods. A fair amount of work can be done underwater, once the proper tools become available, for less cost than bringing the work to the surface, which frequently is the only alternative. An excellent example of this type of problem is related to the pipe lines delivering petroleum products from offshore locations to the beach. Raising enough of a 16-inch pipeline to the surface, even from 60 feet of water, to replace a valve and add a bypass and a connector with a feeder line is a problem that may actually require weeks, providing none of the 16-inch pipeline has other burdens placed upon it, such as competing pipelines across the original installation.

This was recently demonstrated by Transcontinental Gas Pipeline Corporation of Houston when small pipe fittings located in 58 feet of water and attached to a 16-inch line required replacement. The alternatives were underwater repairs or lifting the 16-inch line to the surface and making above-water repair. The 16-inch line was buried and therefore had to be uncovered. It would have to be lifted above the water so that a tap valve could be installed as in the original piping configuration and the 4-inch line reconnected. The break in the line was 2.2 miles from the nearest platform, and two large-diameter pipelines crossed over the 16-inch line just a short distance from the break, making any pickup quite hazardous.

The alternatives themselves were not easy. Cutting and beveling operations, pipe fitting, welding, x-raying, and coding of the repairs would have to be completely carried out under pressure on the floor of the Gulf of Mexico. However, techniques existed and the whole exercise was completed in a few days once the initial technical limitations had been overcome. Saturation techniques were used, with the underwater work site protected by a shell.

3.4 The Commercial Diving and Underwater Mining

Outside the recreational field, industrial and commercial activities include construction of any underwater facility, inspection and maintenance of the facility once it is constructed, survey and design of the use of the underwater site for construction, and the salvage of the underwater construction when it becomes damaged. Commercial diving related to petroleum production is our major underwater activity for man. It can be an exceedingly dangerous activity and one that requires continuing technical advance by the diver to best serve his customer, who is forced to meet increasing costs of exploration and production and increasing competition in yield of his offshore drilling activities. Few facilities exist for extensive diver training. We must start planning.

Some of the early offshore wells yielded the producer 90% of the income and the leasor no more than 10%. These majestic contracts are no longer negotiated. The pattern set recently by the Italian government combine will probably be dominant if more restrictive fiscal ideas do not develop in the next 5 to 10 years.

It is therefore imperative that the cost of exploration and production be reduced dramatically. The cost of diving is high, and many attempts are currently underway to avoid using man in these activities. They have not been totally successful, and some have been complete failures.

Underwater mining represents one of the most magnificently anticipated futures for man if the published projections of the mineral resources are half valid. The alluvial deposits are now accessible. The recovery of the minerals will be a huge industrial effort and will involve man at all his highest skills even if he need not enter the water after the initial phases of geologic exploration or processing development. However, the recovery of mineral resources underwater requires either tunnelling under the seabed itself, as is done in coal mines and tin mines throughout the world, or the recovery of the minerals from the surface of the seabed in varying depths of water.

Such recovery activities can often be paralleled to a huge vacuum cleaner (although dredges are sometimes used), and, as one can imagine, they require large quantities of power. The disturbance to the local ecology, agricultural and fishing activities, if such a vacuum-sweeper mining enterprise is initiated, may cost more than the profit from the mining exercise itself, if deliberate care is not taken to preserve as much of the environment as possible.

The underwater world may require the care and nuturing that we have found necessary on the surface of the earth. Our fish crops this year reached a new low with some species almost eliminated. Planned harvesting, be it of a mining, agriculture, or fishery nature, with replanting, reforestation, and the development of new crops may be an essential portion of any of man's future underwater. The raw exploitation envisioned by some parties will destroy those resources which do exist and are recoverable with the engineering techniques at hand.

A shining example of what may happen if proper plans do not exist from the beginning can be seen in any city subject to the whims of a zoning appeals board that allows spot and strip rezoning which can be observed along the older highways in our nation. The road edges are plastered with automobile dumps, honky-tonk bars, gambling houses, and billboards. A similar result in our underwater environment can be expected from the unplanned, unwise, overenthusiastic attempts to make monkey quickly.

The saturation diving has got many types of limitations. Now, let us try to understand about the problems related to saturation diving.

There are some limitations to saturation diving that must be considered. (The technique is not the panacea for every university or industrial exercise.) One of these is the matter of temperature control. Man lives best in an air environment with rather narrow thermal limits. He is able to tolerate exposures at higher and lower temperatures without losing control of his body heat mechanisms for only limited periods. Then he must have artificial assistance to maintain control of the influence of the environment on his body heat or must return to a more normal environment to recover from the exposure. In water, man loses his body heat for a very obvious reason: water is rarely near body temperature and therefore represents an infinite sink into which man is constantly pumping his body heat. He, unfortunately, does this at a rate that exceeds his ability to produce heat, whether from sustained exercise or the metabolism of food. This means that he will lose control of his conservative exercise and must either leave the hazardous thermal environment or provide an external heat source.

This one problem represents perhaps the most significant limitation of man's ability to work either in a bounce dive or in a saturation dive. This limitation is his inability to stay in the cold water and also conserve his body heat. Informal reports indicate that the majority of the *Sealab II* divers, who had the best available exposure suits and some of the best available thermal protection suits with self-contained heating systems, could rarely exceed 45 minutes exposure in water that varied between 44 and 56 degrees F.

If the habitat used in the saturation exposure is located underwater, as it is in the *Sealab* experiments and the *Conshelf* studies of Cousteau, then the quantity of heat needed to provide both the temperature for the habitat and for the diver is tremendous, and the power load is of some considerable significance. If the habitat, however, is located at the surface and the diver is transferred from the work site to the habitat, via the SDC, the power requirements may not be nearly as severe. If one anticipates self-contained systems strategically located around the ocean bottom, logistically independent of the surface, then power sources of some magnitude must be designed.

A second significant problem in saturation diving is one with which we have had very little experience. We have learned from studies of the closed ecology, as exemplified by the nuclear-powered submarine, particularly those used in the Polaris Fleet, that a considerable number of atmospheric contaminants exist, many of which are not yet identified. Where these contaminants come from remains an unsolved problem.

We describe them all as contaminants because we do not believe that they should be in the gas environment that man breathes and that they do not serve any physiologic function useful to man. Therefore, ideally, they should not be present. With a perfectly designed atmosphere control system they would be completely removed.

It is necessary to design new methods of controlling the purity of the environment under pressure, since we have neither unlimited gas resources nor can we ventilate the habitat indefinitely with compressed gases, regardless of their source.

The reason for this concern about unknowns is that we have no idea whether or not the substances that have been discovered are hazardous to man. We know in terms of direct feasibility studies of the Polaris Submarine Patrols that no apparent hazard to man has occurred within the limited period of observations of the personnel so exposed. It is conceivable that the hazards may not become evident until many years, perhaps even 10 or 15, after the last exposure. We have as an unfortunate but very clear example of such phenomenon in the development of malignancies in the women who so beautifully painted the radium faces on our watches. Not having adequate information about the possible or probable toxicologic hazard of a single exposure under saturation conditions, we also do not know if a man can safely tolerate a second, third, or fourth exposure.

Animal experiments that may provide this information are currently at the planning stage and will be quite expensive, requiring years to satisfactorily complete. Until we do have the appropriate toxicologic information, we can only continue saturation exposures with the hope that we are causing no harm and without the knowledge that the exposures themselves are toxicologically safe.

3.5 The Underwater Sources

The future for man underwater is in the search for and development of freshwater sources, underwater freshwater souces.

The limited freshwater resources throughout the world are becoming progressively embarrassed. We have a choice of either locating new sources of fresh water or of developing desalination progrms that currently require excessive quantities of power.

This may appear to be a minor part of man's underwater activities in the future but can be economically and sociologically significant. It may become necessary for nations to spend a considerable portion of their resources on the development of freshwater supplies for suvival.

There are some secondary problems with desalination. A 150-million-gallon nuclear-powered plant may produce 23,000 tons of salt per day; where shall we put it?

The area of future activities for man underwater may lie in commercial fishing. The average commercial fisherman operating from a power craft spends 70% to 80% of his time searching for the fish that he wishes to harvest. The cost of using man, regardless of the operating cost of the vehicle that man is using, in the United States is excessively related to the market value of his hourly production.

Two possibilities exist that may improve this, both of which provide a considerable future for man in the marine and underwater environment. One is improved search methods, possibly automated, to allow man to reach the crop in a shorter period, to enhance the crop production at less cost and with less effort, making the individual's activities more profitable per unit time involved.

The Environmental Sciences Services Administration has proposed that satellites be used to search for fish. Although this is realistic, it still requires the investment of a considerable amount of money by the Federal Government to create the satellite system and by the fishing industry to develop the communication network necessary to take advantage of the information that may become available from the satellites.

A totally different approach is the deliberate cultivation of the fish crop. (To parallel the attempt of others to coin new terms, we can call this *aquaculture.*) This is now done with shrimp, crawfish, catfish, and oysters and is planned with clams, pompano, and mussels. It is anticipated with lobsters, and it is being tried in a number of cases on an exploratory level with other fish.

Two approaches have been used, one is to take advantage of the saltwater environment itself by creating marine pens to contain the fish. An alternative method uses artificial ponds on land with saltwater pumping systems. Supplementary heat is usually required.

Catfish farmers in Louisiana are producing crops valued at $500 an acre per year on investments that can be amortized and totally depreciated within eight years. This significantly contrasts with the usual corn production in the United States, with the cash yield per acre rarely over $250. Considering the need that we have at the present time for protein sources throughout the world, this appears to be a very successful agriculture endeavor.

A current difficulty, as I personally see it, is that the efforts are in the United States where we do not have a need for increased production of protein. They are not in the areas of the world where protein deprivation is one of the most serious and continuous health hazards the local population has to face. A redistribution of our technologic efforts may represent a profitable aspect of man's future underwater. The cultivation of fish might be an appropriate activity for AID to explore and may actually be far more rewarding than the attempt to develop cattle or sheep in areas where neither animal has previously been cultivated.

There are a number of plants that grow best in saltwater and several that grow excellently in fresh water that are of commercial significance. Their harvest may be a future activity for man in, on, and under the water.

One plant already attracting much attention is kelp. It is an excellent source of protein and might conceivably be used as a total substitute for animal protein, although complete nutritional information does not exist. Presently, kelp is used as a source of a number of products, primarily related to the food and cosmetics industries in the United States. The products from kelp are used as additives to food rather than as a primary food source.

The deliberate farming of kelp in some communities is a profitable exercise. Kelp fishing off the coast of California is now a multimillion dollar activity based entirely on surfaced harvesters of the giant kelp. In the Orient, kelp is used as a food in addition to its other uses, but this is only one example of what may be done. Specific farming and harvesting has not yet been tried commercially in the United States but has been brought to fine art in Japan. The techniques are neither difficult nor expensive but do require skilled labor and reasonable weather conditions. When farmed throughout the world, it may be limited by the labor cost, although mechanization might easily reduce the labor costs.

One contribution of kelp cultivation has gone unnoticed. All of you are familiar with the agar plates that are used to grow organisms in the medical and scientific laboratories, but many of you may not appreciate that the neatness and smoothness in uniformity of ice cream is based upon the kelp harvested off the coast of California, and some of the world-famous desserts frequently attributed to French wisdom are based on kelp called *Irish moss.*

Man's underwater participation in aquaculture will be related to the development of the farm site after its selection, which will be a result of exploration and surveying, and the inspection of the site after the crop has been initiated. Pharmacologic or chemical treatment of the crop during the course of its growth may require man's underwater activities. Harvesting and/or inspection of the consequences of harvesting for the development of the cycle of planting will also require man underwater.

Much of this work could be done from a small, self-propelled submersible vehicle which man could operate at one atmosphere environment. But, as with all other processes requiring direct inspection and/or sample taking, man must, with the present skills available, enter the ambient environment and therefore undergo the hyperbaric exposure.

One other possibility of aquaculture must be mentioned. Another application of plant life from marine sources may be within the closed ecologic space required by man for survival in abnormal environments. Several have proposed the probability of developing a biologic ecology using algae in large tanks to accomplish two things: first, to convert the waste product of man, carbon dioxide, to oxygen which will be consumed by man; and second, to produce, through the regenerative processes of the algae colony, a food source for man.

3.6 Underwater Drilling

In the areas of drilling and well completion for both petroleum and gas, man is used in submersed operations for parts of the exploration, for parts of the construction of the test drill platform, for completion of the well itself, for removal of the test platform, for the insulation of the production platform, occasionally for laying pipe to carry produce from the well to the beach, and for inspection of all underwater structures and equipment. Inspection is necessary to assure that their continued operation will be economically satisfactory and that no danger occurs with respect to the continued production of the actual resource for man's consumption.

The majority of offshore divers are used in the petroleum production industry and, unfortunately, the majority of the accidents that occur to men involved in submersed professional and commercial operations occur in the same offshore drilling industry. This relationship, however, appears to be related to the number of people involved rather than to the inherent hazards of the occupation, although direct statistical information does not exist. Unfortunately, we do not have any data on the number of divers used by this industry, or the number of dives that the divers make, or the number of accidents that do not result in death. And, in fact, we have very poor information about the number of deaths that occur.

There are several reasons for this glaring void in the statistics of occupational medicine. They relate to accumulative errors on the part of almost every individual involved in the industry, but unfortunately reflect on the physicians who are faced with a body and no information in the coroner's office. In addition to many other problems that we have, most coroners in the United States are not trained in medicine; and if they are, they are not trained in the special art and skill for forensic medicine, which is that branch related to the discovery of accidental and criminal death processes.

The common practice in the United States and in many other locations throughout the world is to lay pipeline from the production site (an offshore production platform that is a collecting point of 10 to 60 wells) to the beach and its facilities. Much pipeline can be laid from a barge without the use of divers, as was done recently in the Persian Gulf. Part of that pipe was at depths exceeding 300 feet. However, divers are frequently used in the process of digging a trench to contain the pipe, and then used in inspecting the pipe or placing valves to tie in feeder lines from other and usually smaller production lines or production platforms.

Recently a pipeline project called the Red Snapper Program was envisioned for laying a pipe about 50 miles offshore. The pipeline was to be over 200 miles long with two feeders from this large collection pipe into the beach. The anticipation was that as production wells entered deeper water, the feeder pipes would be laid into the original offshore collection system and that additional feeder pipes from the offshore collection systems at the beach would also be laid.

Under ideal conditions, underwater pipe can be laid in almost any depth of water at several miles per day. However, conditions are not always ideal, and the speed with which the pipeline is laid may be directly related to the conditions at the ocean bottom.

For a pipe to be reasonably secure, it must rest on the bottom, preferably in a trench. A pipe suspended between two high points of the ocean bottom becomes a swinging limb in catenary shape and rapidly disintegrates due to water movements and due to the probability of its colliding with objects moving with the currents.

If one considers the activities on the Gulf Coast where completion wells are now going in at 250 to 350 feet of water and where divers are used consistently and repetitively in these operations, and if one accepts the projections made by the American Association of Oil Well Drillers that untold resources exist on the Gulf Coast out to water depths at least to 1500 feet, he can envision the complexity of the operation for the diver and the magnitude of the employment possible. The Santa Barbara Channel, with 1,200-foot depths, has just opened and increases the task.

Needless to say, the cost of commercial diving and the accident rates, regardless of their causes, are such that a number of the drilling corporations and their employers who own the leases are rapidly searching for and trying to perfect methods of exploration, well drilling, well completion, and production completion that do not require a diver. While it is not our hope that divers will become unemployed, it is our hope that methods will be developed that prevent accidents. If wells can be completed satisfactorily at the production level without the use of a diver, then the industry has improved significantly.

Whenever a well is completed and becomes a production facility, it is necessary to deliver the produce to the beach for consumption. There are two ways of doing this. The first requires the use of a tanker or barge system to move the produce from the production site to the beach for off-loading into some other facility. This usually requires that some storage facilities exist at the production site offshore to make it economic to use a tanker or barge system.

On some parts of the Gulf Coast, the maze of pipes is so complex that new pipe handling and pipe repair techniques have been required. You will remember the Ocean Systems, Inc., demonstration of placing a hot tap on an underwater pipe. That operation was necessary, principally, because the pipe that needed the repair in the hot tap was already overlaid with competitors. Lifting the one pipe section needing repair would have been an expensive, as well as tremendously hazardous, operation.

These continued uses of man in submersed operations should increase in number. They represent one additional reason why good selection, good training, and good qualification programs for underwater workers will be needed. It is obvious that as the complexity of the submersed operations increase, more than divers will be needed. Skilled technicians and skilled construction personnel are required. The training for these skills is longer than the training for diving. Thus it will be necessary to find construction personnel, geologists, welders etc. and to train them for work in the water environment under high pressure.

This requirement for highly skilled personnel to work in submersed operations raises another interesting problem. We have no training facilities, that I am familiar with, that satisfactorily train men or women for commercial diving. In making this statement, I am quoting my friends in the professional diving contracting industry.

Our past inabilities to safely enter the water, stay there for any useful period of time, and leave it have been sharply restricted by the thermal conditions of the exposure, the improbability of safe decompression information, and the limited sources of compressed gas required for breathing media. It was not until the turn of the century that even a basic theory of decompression existed which allowed the prediction and calculation of probably useful tables from moderate depths of exposure. The problems of thermal exposure have not yet been solved in any reasonable fashion, short of pumping hot water continually through the diving suit.

Sources of good, reliable, pure compressed gases are still limited throughout the world and may, for many of the activities described in the previous paragraphs, represent, the economic or financial limitation of the enterprise. For large underwater operations, liquid gas may be part of the answer.

However, if surface power sources are available, gas-recovery compression systems could be used to provide the underwater breathing media. A surface supplemental heat supply would be secondarily available for heating the tethered diver. The required supply of heat for the free diver, using scuba, is still currently limited, although a number of feasibility studies have been done and a large number of proposals exist for the use of isotope heat sources to provide hot water for circulation within the diver's dress.

The prime market of the oceans will be the things produced from the sea. The hardware, the systems, the materials, and the services needed to support production from the sea are ancillary markets and will develop slowly. Corporate appreciation of this difference will stimulate the large investments needed to develop the produce for the primary market. Private activities, rather than governmental, will be the major stimulus to the economic aspects of the hydrosphere.

A single example may be projected in the area of the drug industry. The use of marine biologic systems to provide needed or new substances is anticipated. The use of the same systems to control the production of desired end products is entirely exploratory. Aquaculture may be controlled by drugs, marine minerals dissolved in the water may be concentrated by controlled biologic systems, and the proteins remaining after appropriate extraction have a ready market.

The legal problems related to man's future underwater need clarification. This will be a slow evolution because agreement does not exist in problem definition. It will be only after considerable soul-searching that the needs of a new law-of-the-sea will emerge to stimulate the exploration of the hydrosphere, thereby enhancing man's future underwater. The slow progress will relate to the market for the produce and the cost of the risk investment, but the progress is inevitable and the future is potentially magnificent.

Chapter Four

The Oceanic Biomedical Sciences

4.1 Introduction

The needs in oceanic biomedical sciences are similar to those of other sciences that have peculiarities related to the environment to be used and studied.

The biomedical sciences combine biology and medicine for the study of complex systems, including the relationships between man and the abnormal environments that he may encounter while attempting underwater work or research. These will include man's selection, training, qualifications, educational requirements, fields of interest, performance, psychological fitness, and exposure to hazards of the environment, the hazards of the technical crutches provided for him and hazards within the environment, like poisonous and venomous fishes, and other forms of biological phenomena, creatures etc. that may be unexpectedly encountered.

Educational needs are not only those of the individual and his associates but also the needs of the community. The educational needs will be determined by what we believe to be man's future underwater.

In order to establish an educational program, one must anticipate the people that will be needed, the services required to support the people, and the purpose of the entire program with respect to its benefit to the community. Physical facilities will be needed beyond those we have. Even though a number of colleges and universities have established programs related to oceanic sciences, engineering and marine biology, a program does not exist which encompasses all the oceanic biomedical sciences at either one facility or university or at a combination of them.

Schools, laboratories, and research facilities in some instances may be isolated from the oceanic environment. Some must obviously be at least on the water's edge or surface, permitting modest access to the water, be it river, lake, or ocean. Ships, piers, wharfs, and various fixed platforms are all required.

The tremendous advantage of being able to work in the environment one wishes to study or study the use of can now be appreciated after the *Conshelf* and *Sealab* experiments. As soon as ambient pressure work is accepted, the choice then remains whether or not to initiate a saturation program. Laboratories in research institutes with saturation facilities could have an advantage in some areas, although for many years most scientific activities will be accomplished quite satisfactorily without them.

The vehicles themselves permit mobility from the point of water entry in terms of radial distances as well as mobility in depth. For many years the pure exploration of water environment, which has never really been seen before, will be one of the major occupations of vehicles. They can and will be used in the search and recovery of misplaced articles. The pure exploration will be the exciting activity.

All this work, unfortunately, will develop slowly because the hazards of the environment itself and the problems of sea state and weather control are such that facilities will be very expensive to construct and to operate. For this reason, rapid communication among investigators will be needed to conserve the limited resources. Data handling centers will be required to assimilate and control and distribute through networks and to libraries the immense amount of new information that will be provided every time anyone enters the water.

There are several types of activities that one can see in the future in biomedical sciences in the oceanic environment. They might be organized as follows: some activities will be mission-oriented, some will be geographically- or regionally-oriented; others will be service-oriented to support perhaps the first two activities. A fourth category will be that of exploration. Advancing from this will be the study of feasibility and/or testing to establish standards that will influence in turn the mission-oriented, the geographically-oriented, and the service-oriented activities.

Whenever man is going to enter the environment, there is a need for the physician, expert in diving medicine, the toxicologist, the pharmacologist, the physiologist, and the psychologist. The scientists who might study in the environment would be the biologists themselves, the botanists, fishing experts, kelp growers, biochemists, and physical chemists. In order to understand the environment that man is in, geologists are needed as well as data collectors, photographers, librarians and data processors, and experts in navigation and optics.

If the habitat will be down for very long and will be relatively independent of the surface, independent power systems like reactors will be required. A new group of support and service personnel will have to be maintained to provide the proper facilities and to provide the means and manner to safely operate the facility for the support of the work site or the research tools others wish to use.

You will notice that many of the titles used in describing people are titles which usually refer to college graduates. This is a matter of semantic simplicity rather than a matter of educational requirements. If any of you have done field work or laboratory research, you know that not all participants in every phase of every study need be college graduates. Certainly they do not need to have a doctorate.

Very few of us are able to work unassisted by technicians and technologists, instrument personnel, prop people etc. They must all be trained in their technical areas so that they become the hands, eyes, and ears of the principal investigators.

Let me give you an example of an operation working from a barge and involving decompression diving with a scientist as the diver. To do this safely, one must have a standby diver, who need not be a scientist. One also needs tenders and a diving supervisor. The latter must be qualified in all diving operations and have absolute control of the diving activity. Since the diving will be done from a platform of some sort, if it's mobile, it will have a crew and the crew will have a leader. If the diving is done to any depth, there will be recompression chambers, chamber operators, people to do gas analysis, medical assistants of various levels, data handlers, time keepers, photographers, communicators, and others. Many of these need not be college graduates, and, in fact, it would be wasteful to use them in this fashion.

Or consider a marine biology program requiring a principal investigator, scientific associates, technical assistants, data handlers, clerical assistants, supply and logistic people, a supervisor at the working surface, sample collectors and the vehicles they use, pilots, life-support people, diving supervisors, medical assistants, chamber operators, and communications and navigation experts.

The use of habitats will require even more personnel, technically qualified and well-trained, but not necessarily university people. This means that we must rapidly create the facilities needed for training these technicians and technologists, regardless of their ultimate skills.

Most of the people who support our activities at the present time either receive some training in the Navy or receive training as the result of on-the-job activities. They enter into their employment as absolute novices and slowly improve in their skills and understanding until they become quite technically competent.

Scientists, skilled in the biomedical fields, will be needed to enter the oceans to apply their special talents. Many may not and, in fact, will not have any contact with the oceans until most of their training is completed. We have already developed inland schools giving courses which lead to degrees in oceanography and marine sciences.

An oceanographer is generally thought of as a physical scientist with several degrees. In most instances he is required to obtain a Bachelor of Science degree and to do additional work at the graduate level. After two to four years he may earn a master's in oceanography and after another four years or so a doctorate, and then, if he is still ambitious, he can return for additional postgraduate work.

These rigorous academic standards are not required for most of the people who will be needed, and we should not let descriptions of the training programs for advanced degrees inhibit the interests of the young people in the country who become excited by the oceans and the studies that can be done on them and in them.

4.2 The Biomedical Technicians

In the biomedical fields technicians can be trained in high school. Most laboratory work including data handling requires one or more assistants for the principal investigator.

We must, to provide ourselves with useful, technically qualified assistants, find them early, train them quickly, and provide them with the incentives and desires to do progressively better and more exacting work.

How many do we need? Current estimates, based on the experience of established laboratories devoted to environmental studies, are that one senior investigator uses five people, not counting housekeeping and library personnel. In a modern hospital each physician requires about six people to support his twenty-four-hour, seven-day-a-week program of patient care. This means that for each physician graduated we need to technically train, at some level, six more people, and for each Ph.D., assuming he becomes a senior investigator, five more people. Most medical schools fail to do this. We still abhor the shortage of semiprofessional people. Our technician training program should be at least four times the current size to meet the needs of the community and provide hopes of overcoming the current deficiencies.

Intermediate between the technician just described and the professional required to design and supervise the biomedical studies of oceanic problems is another group deserving special education: the technologists. These are the people who have been trained beyond high school in college–level facilities for one or more years in relatively narrow areas with advanced technical experience in the laboratories. Their curriculum would perhaps be more time-consuming than the curriculum of the professional student and the time spent on learning technical details more extensive.

These people, now almost completely unavailable, will be the strong right hand of the principal investigator. They will set up experiments, complete the logistics, arrange for data handling, and assure that the required equipment is available, presented, and properly manned by suitable technicians. Many senior scientists have, after years of effort, acquired such a technologist, usually from his own laboratory after years of on-the-job training. But we need thousands of these technologists in the oceanic biomedical sciences now.

Duke University has initiated a program for a physician's assistant, but this is oriented towards clinical activity and not laboratory efforts. Purdue University has technologist programs in several engineering disciplines but not in any biomedical fields.

It will be necessary for our medical schools and our graduate schools to design the job descriptions and describe the intellectual achievements required of the future technologists. Then they must convince those who control the finances of the educational systems of the acute need for this specially educated and trained person.

The educational need in the oceanic bio-medical sciences is for persons who can speak two or more scientific languages. These will be people with solid training in both physical and biological sciences. Some possible examples are anatomy and mechanical engineering, microbiology and mathematics, communications and physiology, medicine and fluid dynamics, electronics and neurology, thermodynamics and molecular biology.

Cross training like this has been proposed and has been started in several universities.

The new *discipline* is often called *bioengineering*. It is dignified by at least four English language journals, some national and international societies, and a number of international meetings. Norbert Wiener once called it *cybernetics*. The US Air Force calls it *bionics* and sponsors annual conferences for its own contractors.

Why do we need these people in oceanic biomedical sciences? Because the work is in the water and some of it is under unusual pressure. All work is done in an environment considered abnormal when contrasted to our terrestrial experience. Even the collection of specimens may be difficult.

The hyperbaric exposures create new problems in toxicology, pharmacology, physiology, pathology, and biochemistry for the man during his own exposure. The control and use of special pressure-compensated equipment, the design and use of hyperbaric chambers, and the ability to usefully manipulate the extraordinary environmental physical features in biomedical interests demand more skills, more training, and more understanding of the physical sciences than is the common fortune of the biologist.

Here at the University of Hawaii there is an unusual opportunity to provide just this cross-training needed. If and when the Medical School expands its efforts from the current preclinical program to include appropriate clinical training, the potential stage will be complete. Unfortunately, as long as the medical student must leave the Islands to complete his training, it will be difficult to provide the continuous cross-training needed.

Some changes might be anticipated in the various curricula of the many colleges and schools within the university to permit students in these programs to take advanced-level courses without all the usual prerequisites. This latter need may actually force a reexamination of the current courses themselves since some are undoubtedly bound more by tradition than by current needs.

4.3 The Decompression Injuries

Decompression injuries are defined as those injuries to man that result from some event during or after decompression from a hyperbaric exposure and that are considered relatable to the tissue content of inert gas.

An obvious accidental injury occurs when the diver overinflates his dress, and, because of hyperbuoyancy, *blows up* to the surface. He might not be only physically injured by collision but might also be quite vulnerable to injury due to unequal distribution of gases in body cavities and to bubble formation. His susceptibility to gas embolism is increased if his lungs were not adequately vented during the rapid ascent[1]. These are injuries of some considerable concern in operation conditions. (Many features of gas emboli have been well-described and will not be further amplified.)

The more pressing problem is one involving tissue gas tension, or classical decompression injury. Decompression injuries are considered the consequence of inadequate decompression following hyperbaric exposure. Similar problems follow rapid ascent to altitude, and many features are comparable.

The mechanism of symptom production is believed to be bubble formation within the tissues as a result of the differential occurring between the gas tension in the tissues and the hydrostatic pressure of the environment. Under a diving pressure, the gases, based on their solubility, are dissolved in the tissue fluids. When this pressure is dissipated, the probability that gases in solution will revert to their physical gaseous state, while within the tissues, is high. (Experimental production of tissue bubbles is easy.)

The location of the bubbles that produce symptoms remains unresolved. A few authors believe that the bubbles are entirely intravascular; but most believe they are entirely extravascular. The rate of decompression that permits the production of bubbles may actually determine the physical location of the bubbles, but the relative significance of intravascular *versus* extravascular bubbles remains to be determined. For brevity it is necessary to exclude the problems related to the pulmonary hyperexpansion disabilities commonly associated with the phenomena of gas embolism.

If one assumes that the bubbles are extravascular, then it is necessary to determine whether those physiologically and pathologically significant bubbles are inside or outside the cells of concern. Clinicopathologic evidence from nondecompression processes associated with the accumulation of masses in soft tissue suggests that intracellular compression of pain-sensitive structures is relatively unimportant but that extracellular masses distorting the physical characteristics and dimensions of the normal structure would be significant and are painful.

An immediate clinical point of comparison is the difference in probability of pain, as a significant symptom, following trauma, which produces considerable edema, versus one of the usually painless storage disorders in which the accumulated material is contained within the cells. Several of the storage disorders with a slow time course, with accumulations in the peripheral nerves, relatively near the nerve endings, are not characterized by pain as a significant symptom. It is doubtful that intracellular, neuronal bubbles are significant in the production of pain in decompression injuries.

This, however, is a relatively minor point and not of particular theoretical concern at the present time. If we assume that any symptom production is related to the problem of bubble formation, it is then necessary to explore the aspects related to the production of bubbles, regardless of their location.

Human tissue is normally saturated with gases at pressure of 1 ata, the pressure of the environment in which we usually live. People on mountain tops have their tissue saturated at a lesser pressure. When one pemits exposure to a hyperbaric environment, the increase in pressure acts as a force pushing gases into tissues no longer saturated.

The capacity of a specific tissue for a particular gas is based upon the gas solubility within the tissue, the diffusion rate of the gas through the particular tissue, and the tissue perfusion. The perfusion of tissue, regardless of gas solubility or diffusion rates, is of some considerable importance because gas cannot enter tissue any faster than it is delivered to the tissue.

It is therefore evident that tissues with a high gas solubility, with high diffusion rates, and excellent perfusion will accumulate gas under a pressure differential most rapidly. It is also assumed that this same tissue will lose the gas faster under the conditions of decompression (reduction of pressure) than tissues with slower diffusion rates, lesser solubilities, or less perfusion.

Under these assumptions, calculation of decompression tables is currently based on saturation rates of hypothetical tissue compartments. On an empirical basis this technique has been found to be relatively satisfactory. This, in itself, is absolutely no evidence at all that the assumptions are correct. The lack of theoretical validity against the practicality of a successful, nonsymptom-production decompression schedule is relatively unimportant operationally but is of concern when designing new decompression tables.

A major limitation in all diving operations is the costliness of the decompression time compared to the operating time at the work site. According to U.S. Navy Decompression Tables, a 30-minute dive to 300 feet on air requires almost three hours of decompression time. These tables have been calculated on the assumptions that perfusion is the limiting factor in determining the tissue gas content and that an arbitrary number of tissue saturation rates (half-times) exist for calculation. These assumptions have been empirically useful for the past 13 or more years. Unfortunately, many attempts to save decompression time have resulted in schedules that produce decompression injuries. Fortunately for the diver exposed, the injury may be relatively minor although frequently quite painful.

Most current theories of decompression injury predict a *metastable limit* relative to the supersaturation of tissues prior to the production of symptoms that are the result of bubble formation. This may not necessarily be true. It is probable that random nucleation of gas phase separation occurs within tissues, particularly if one considers the probability of cavitation of liquid-liquid interfaces which provide reasonable theoretical models for tissue studies or the random nature of symptom occurrence.

The concept of supersaturation of tissues in a reliable metastable state allows relatively satisfactory calculation of decompression tables from hyperbaric exposures and will permit symptom free surfacing of most individuals exposed. It is assumed that one can surface with a specific overload of supersaturation in each tissue compartment with safety and that the quantity of saturation of the various tissues is related to their half-times. Unfortunately, our experimental skills are too limited to test whether a tissue can be fully saturated or whether tissues may also have, under uniform conditions, varying but reliable half-times for saturation.

However, if one mathematically expands the concept from a finite number of tissue compartments to an infinite number, no apparent added advantage in the calculation of decompression schedules occurs. This suggests that there may be some principle and/or physiologic factor over looked among the basic assumptions made for the calculation method.

4.4 The Physical Process of Decompression

The significant component of the physical process of decompression is the composition of the gas inhaled during the exposure and decompression.

Tissue saturation with an inert gas, like nitrogen or helium or neon (if one can afford it), is proportional to the duration of the exposure and the partial pressure of the gas to which exposed. The desaturation of the tissues is similarly related. The time spent in the process and the gas partial pressure difference between tissue and ambient environment determine the tissue desaturation rate. Therefore, variations in the gas inhaled both during exposure and during decompression are significant.

It is current diving practice when using helium exposures to include oxygen breathing as part of the normal decompression schedule. By completely eliminating helium from the breathing gas during the process of decompression, one maximizes the pressure differential between the tissues and the environment, greatly enhancing desaturation of tissues and therefore theoretically reducing the probability of symptom production.

Keller and Bühlmann have taken advantage of this concept and have successfully performed a number of dives. By changing the gas composition during both compression and decompression, they were able to use decompression schedules of shorter duration than would normally be considered permissible. They completely switched the inert gas during the process, usually more than once. They also took maximum advantage of the use of oxygen and produced some remarkable decompression schedules.

However, it must be noted that not all of the diving programs have been successful, and that many of their dives produce symptoms. These are not described completely, however, and it is hard to decide whether the symptoms are produced as a consequence of overextending the probable limits of the design.

Additional problems that may influence the probability and/or susceptibility to decompression sickness are the toxicologic hazards of oxygen and the pharmacologic properties of nitrogen or other inert gases. Oxygen poisoning is well-known and can be avoided only by physically avoiding exposure to a toxic environment. This requires, with short exposures, that the oxygen partial pressure be kept at least below 1.5 atm, and for prolonged experiments of several days, perhaps even less than ½ atm.

The inert gas narcosis problem, characterized by nitrogen, is emphasized by exposure to gases of high molecular weight. The model is best characterized by the use of xenon by Featherstone and his coworkers as a general anesthetic at 1 atm with gas mixtures of 80% xenon and 20% oxygen. The necessity of avoiding narcosis due to the nitrogen requires that helium be used as an oxygen diluent.

Other gases should be considered because of some difficulties in the use of helium. Helium has a high thermal conductivity and presents a problem in conserving body heat of the exposed diver. It also, because of its physical characteristics, distorts speech dramatically as would probably hydrogen.

The effective use of decompression schedules is influenced by exercise during either the exposure or the decompression period. Too much exercise may increase the production of symptoms of decompression sickness. One can assume that exercise accomplishes a number of things.

First, it would probably increase the perfusion of tissues because the cardiac output would be increased and more blood would flow through the vessels in any unit of time. This would either increase the delivery of gas to the tissues or, if one were in decompression, enhance the removal of gases from the tissue-vascular bed interface.

Simultaneously, exercise will increase oxygen tension within the tissues, which may alter the diffusion etc. of gases through the tissues, and will, because of the basic metabolic requirements of any biologic system, produce more carbon dioxide within the tissues. The carbon dioxide has a very high solubility in tissues and in blood and therefore can diffuse quite readily. Under appropriate circumstances carbon dioxide can diffuse into a preexistent, silent bubble, causing its expansion. This may enhance the production of significant bubbles and may actually, although never proven, after hyperbaric exposures, increase the possibility of symptom production.

In addition, exercise requires that tissues be moved. Whenever relatively rigid structures move with respect to fluid interfaces, cavitation is more likely. The increased probability of cavitation under these circumstances, when the tissues are saturated with gases in solution, is such that exercises resulting in increased perfusion should enhance removal of gas from the tissue; but exercise, because of increased distortion of extravascular tissue may also enhance the probability of bubble formation by cavitation.

It appears from a number of tests both with altitude and following hyperbaric exposure that exercise during decompression is a harmful procedure and that exercise dramatically increases the probability of symptom production. Remember Dr. Miles' statement that rest is an important feature in the avoidance of decompression injuries.

The external influences discussed above affect the probability that an individual will become more or less susceptible to decompression injury. Many are of physiologic or medical import and difficult to discuss without clinical experience. We fail to understand the mechanisms of most of the external influences because of inadequate data and therefore do not appreciate the specific parameters involved. It is the hope that, from the *Sealab* and the *Conshelf* exercises, data are collected that will provide the proper direction to our underfunded universities and nonprofit institutions in their own endeavors.

Another problem of some considerable theoretical value that has interested many for years is why decompression injuries occur one day and not the next. Some philosophy must be mentioned before delving into the physical problem.

The biologic system, which is the envelope of the process characterized by symptom production, is considered in all methods of calculation for decompression tables as a closed thermodynamic system. The assumptions used in the design of the mathematic schemes imply no alteration in entropy. The equations used require that each individual have identical characteristics at all times and ages. However, there appears to be an increase in symptom production related to age, time of day, meals, or temperature changes of the environment. These variations are not consistent with closed systems.

If we assume that symptoms and injuries are the consequence of bubble growth, the mechanism of bubble formation within the biologic system must be determined. This should include, of course, the reasons for the variations in biologic susceptibility.

Bubble formation has been a topic of interest ever since economy was introduced into industrial operations. Bubble formation became an academic interest and papers have appeared in the theoretical literature as well as in the applied sciences publications.

During decompression, bubble formation within biologic systems is considered to occur by one of two mechanisms. Some will argue with the number of mechanisms, but most propositions may be so reduced.

It is assumed that some fluids of the biologic system, regardless of their location, will be super-saturated to some degree during or after any compression experience when the exposure is short. Since more gas will exist within the fluid compartment than could normally be contained in a uniformly stirred solution under standard conditions, the probability for phase separation is high.

It is further assumed that phase separation has not yet occurred, in the absence of symptom production. Even though excess gas is in the fluids, a metastable state exists, highly susceptible to bubble formation with the slightest physical disturbance.

It is also believe, within this hypothesis, that bubbles are initiated by cavitation. Once the cavity is formed, if the pressure differentials are sufficient, gas will enter the cavity, forming a bubble. The probability of the latter event is related to the surface tension of the initial cavity. If bubble growth occurs, due to the pressure differentials, symptoms will occur when a bubble of sufficient size is appropriately placed.

The relationship between bubble growth under these circumstances and the gas available to permit the bubble to grow, is such that 50 to 100 times the cavity diameter would need to be completely desaturated to provide the gas mass required to overcome the initial physical constraints on growth.

Deformation pressures develop with time. Most calculation methods require that surface dimensions remain constant and ignore any transient variations in elastic properties of the tissue. Usually when cavity pressure increases with respect to surrounding hydrostatic pressure, the vólume and, therefore, the surface dimensions change whenever the cavity environment has elastic properties. Such arguments have weakened the theories to justify current models of decompression needs.

An alteration of thinking is required, although only a few have made bold new proposals. Pease and Blinks discuss *de novo* bubble formation from cavitation in the absence of any preexisting gas phase. They treated their fluid systems according to Harvey and completely confirmed the observations of previous workers that fluid systems contained in glass or rubber could be degased and no applied force would produce cavitation.

They demonstrated, however, that certain types of solid surfaces greatly facilitated easy cavitation in the absence of gas nuclei. They conclude that the cavitation was produced by starting a tear when no gas phase existed. The tensions applied as driving forces were small, and the tear must have started at a very *weak* locus where intermolecular forces momentarily approached zero. *Hydrophobic* surfaces with exposed nonpolar groups allowed tears to start most readily.

If one assumes that a fixed molecule is less likely to follow thermal movements of fluid molecules around it, and if the forces of adhesion are low at the start, such as van der Wall forces, then it might be possible for a fluid molecule to momentarily escape from the attractive sphere of its associated fixed molecule. A hole, therefore, comes into existence.

These were believed to be the conditions when hydrocarbon chains were exposed on solid surfaces. The characteristics of the fluid made little difference. Since chemically identical fluid systems without the solid surface, do not cavitate, the corollary of the above hypothesis is that the holes do not develop in the body of the fluid.

The development of cavities in fluids and gels has been studied and the pressure changes, or energies, required within the *in vitro* models are often of the order of 100 atmospheres. Measurement of such pressures within tissues has not been done, although shear states might conceivably occur that allow transients of these magnitudes. Translational energies applied to a container of gas under pressure differential will frequently initiate bubble formation even after strenuous efforts to completely degas the initial solution. If bubble nuclei (specific gas bubbles) remain in any solution, bubble formation does not occur, but bubble growth controls.

There are several reasons for not immediately accepting these ideas. Bubble nuclei probably never exist in biologic systems. The mechanism of their introduction itself would pose a considerable problem. Exotic solid surfaces are unlikely. Experience with them does not exist in biologic systems, except in the instance of calculi formation, which has yet to be implicated in the problems of decompression injuries. Therefore, additional thinking is needed on the mechanism of bubble formation within biologic systems – the fluids being the simplest state.

CHAPTER FIVE

THE MOORING ROPES

Introduction

The usual standard for moorings have a directional twist opposite the strands themselves. If the rope has clockwise strands the lay is called right hand. It is left hand if the strands are counter-clockwise.

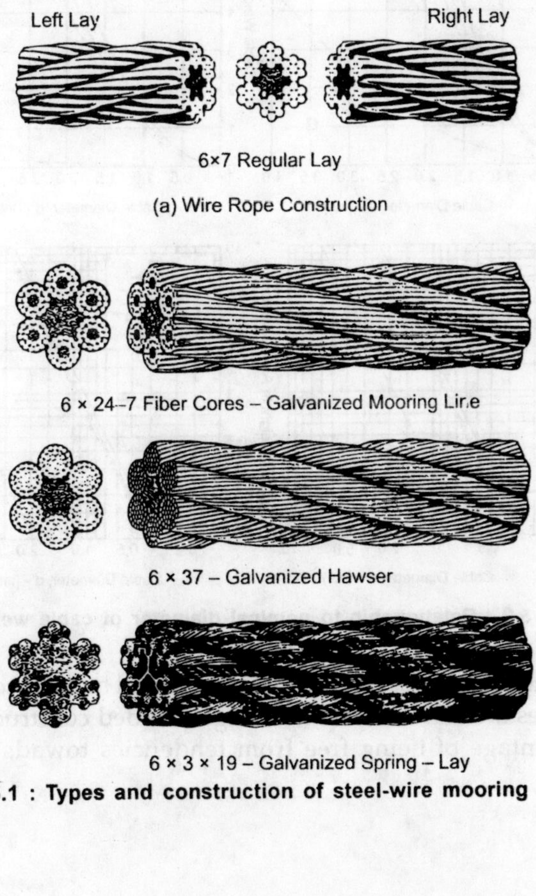

Fig. 5.1 : Types and construction of steel-wire mooring ropes.

The steel in wire ropes may have different grades of strength and may be galvanized against corrosion. A steel wire rope usually comprises six strands, each containing 19, 24 or 37 individual wires built around fiber or steel cores. Fiber ropes, on the other hand, usually consist of three strands, *plain-lay* but may be obtained in four-strand, *shroud-lay*, or nine-strand, *cable-lay*, construction at some sacrifice of strength.

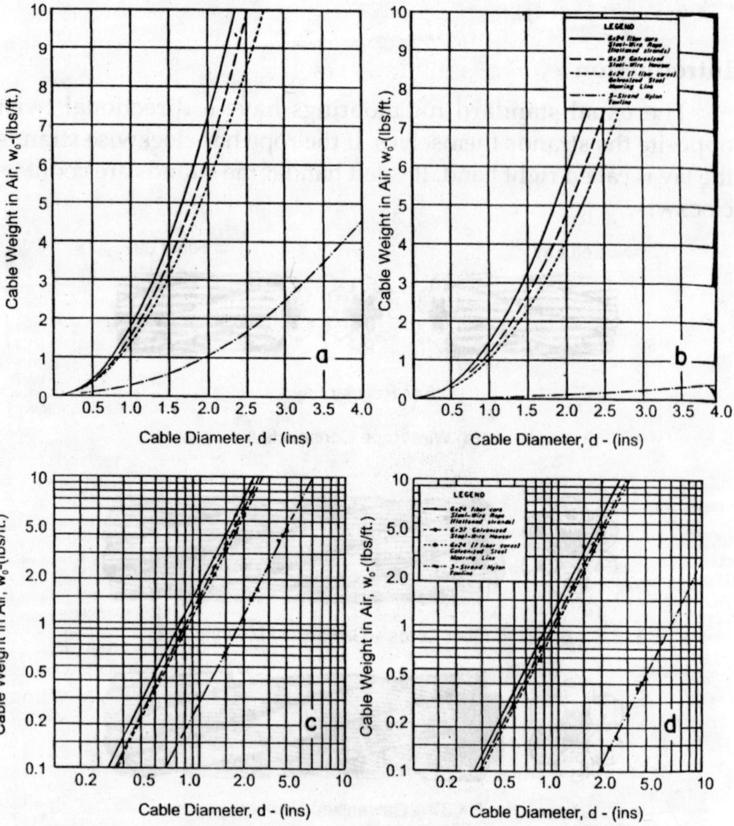

Fig. 5.2 : Relationship to nominal diameter of cable weight in air and water.

Besides layed ropes, plaited fiber ropes and braided core-cover fiber ropes are also used for moorings. Braided construction has the advantage of being free from tendencies towads kinking because half of the 8 strands are left-hand lay and half are right-hand lay. Eight-strand plaited rope has about the same strength as layed fiber rope of the same material. Braided rope gains in strength efficiency from the nature of its construction.

Typical fiber materials are manila, sisal, coir, nylon, saran, dacron, rayon, and prolene (polypropylene). Manila hemp is grown in the Philippines; sisal fiber comes from Mexico, East Africa, and Java, and coir is imported from India. The remaining fibers mentioned are synthetic, manmade products which provide certain desirable properties not found in natural fibers.

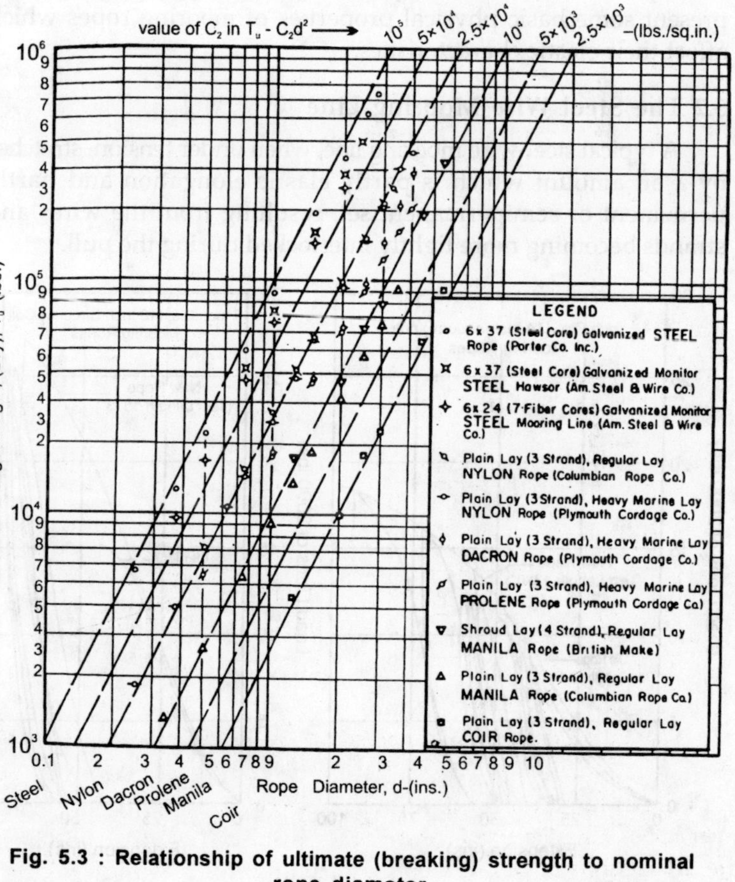

Fig. 5.3 : Relationship of ultimate (breaking) strength to nominal rope diameter.

Under tension, a mooring rope does not have the linear load-extension relationship characteristic of many solid materials. The strands and individual fibers contract, and the stress-strain relationship is nonlinear. If, after loading a mooring line, the load is removed, there will be permanent or semipermanent stretch or *set*, a result of interstitial friction and compaction. When the line is reloaded, the load-extension relationship develops hysteretic loop with the unloading curve. Repeated loadings and unloadings create a succession of hysteretic loops. As a result, the set stabilizes or the rope fails. These effects will be discussed more fully in subsequent sections of this paper. It seems appropriate first to present some basic physical properties of mooring ropes which affect their elastic characteristics.

5.2 The Steel Wire Mooring Line

A typical steel-wire mooring line, when under tension, stretches by a an amount which is partly elastic elongation and partly permanent or semipermanent set, resulting from the wires and strands becoming more tightly interlocked during the pull.

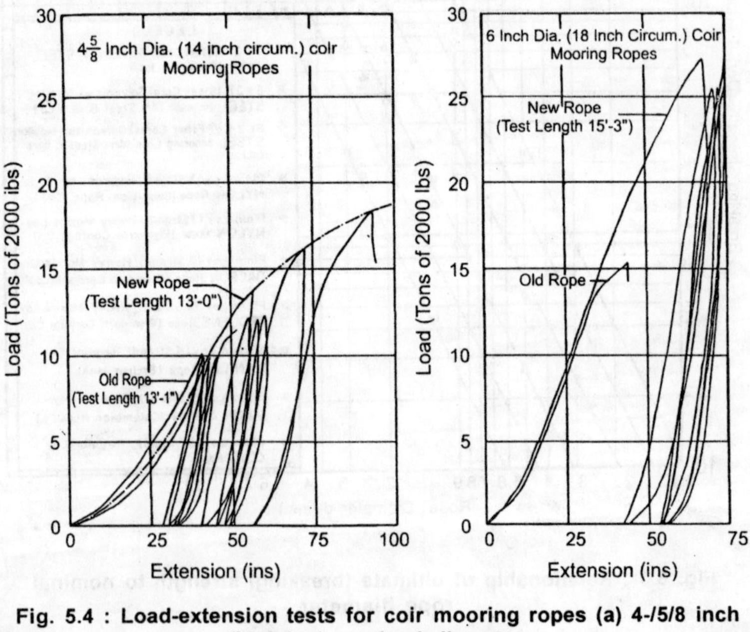

Fig. 5.4 : Load-extension tests for coir mooring ropes (a) 4-/5/8 inch (b) 6 inch nonimal diameter.

If the load is completely removed, the curve of unloading follows a different path on the stress-strain diagram and establishes the amount of permanent set on the axis of abscissae. If the load is reimposed, the new loading curve forms a hysteretic loop with the unloading curve, the area of which represents energy absorbed during the unloading and reloading cycle. In general, if the loading and unloading cycle is repeated, the successive hysteretic loops superimpose on each other and show a slight backward drift (towards greater extension), indicative of a further accumulation of permanent set.

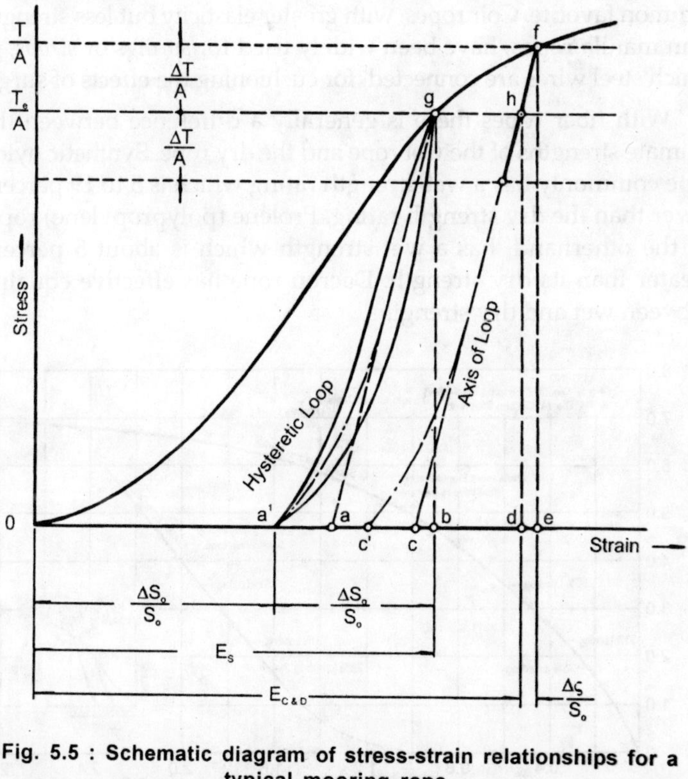

Fig. 5.5 : Schematic diagram of stress-strain relationships for a typical mooring rope.

Under repeated loading conditions, when tension is not entirely removed from the line, the tension-extension relationship is nearly linear along the upper portions (dash lines) of the axes of the hysteretic loops. If, for example, a 1-1/8 inch diameter, steel mooring line is loaded to an equilibrium tension of 25 short tons and then, under repeated loading, varies tension within the range of ±10 tons about the mean.

5.3 The Strength of Ropes

Excluding steel ropes, nylon ropes have the greatest strength for their size and would therefore be of great value for certain types of mooring, particularly on account of their low weight. Dacron, prolene, and manila ropes may be of value when less elasticity is required than that of nylon ropes. Manila and coir ropes, of course, have been used extensively in harbor moorings, manila being a common favorite. Coir ropes, with greater elasticity but less strength than manila ropes, have been widely used for *springs* or *strops*, to which steel wires are connected, for cushioning the effects of surge.

With fiber ropes there is generally a difference between the ultimate strengths of the wet rope and the dry rope. Synthetic nylon rope commonly has a wet-strength rating which is 5 to 19 percent lower than the dry strength rating. Prolene (polypropylene) rope, on the otherhand, has a wet strength which is about 5 percent greater than its dry strength. Dacron rope has effective equality between wet and dry strengths.

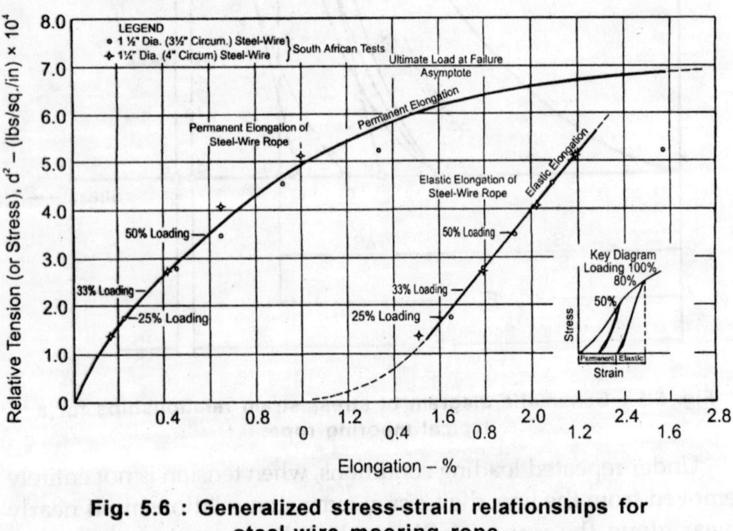

Fig. 5.6 : Generalized stress-strain relationships for steel-wire mooring rope.

Weights of Mooring Cables

Type of Mooring Rope		Constant of Proportionality (lbs/ft/sq in)	
Material	Construction	In air, C_a	In water, C_w
Steel Wire	6 × 24 Fiber Core, flattened strands	1.80	1.45
	6 × 37 Galvanized hawser	1.55	1.20
	6 × 24 (7 fiber core) Galvanized	1.40	1.05
Nylon Rope	3-strand towline	0.29	0.028

Ultimate Tensile Strengths of (New) Mooring Cables

Type of Mooring Rope		Constant of Proportionality (lbs/sq in)	
Material	Construction	In air, C_u	In water, $(C_u)_w$
Steel Wire	6 × 37 (steel core) galvanized steel rope	8.8×10^4	
	6 × 37 (steel core) galvanized monitor steel hawser	7.3×10^4	
	6 × 24 (7 fiber core) galvanized monitor steel rope	6.5×10^4	
Nylon Rope	Plain-lay (3-strand) regular lay	2.3×10^4	$2.1\text{-}2.4 \times 10^4$
	Plain-lay (3-strand) heavy marine lay	2.3×10^4	2.1×10^4
Dacron Rope	Plain-lay (3-strand) heavy marine lay	1.85×10^4	1.85×10^4
Prolene Rope	Plain-lay (3-strand) heavy marine lay	1.80×10^4	1.90×10^4
Manila Rope	Shroud-lay (4-strand) regular lay	1.00×10^4	
	Plain-lay (3-strand) regular lay	9.0×10^3	
Coir Rope	Plain-lay (3-strand) regular lay	2.5×10^3	2.25×10^3

5.4 The Mooring Line Geometry

In the case of harbor moorings the geometry radically differs from other types, and the effect of the moorings on the ship motion can be counted on to be much more dominant.

The mooring lines are usually relatively short and hang in air as noncoplanar catenaries, while the ship, in shallow water, may be subject to considerable change in elevation from alteration of

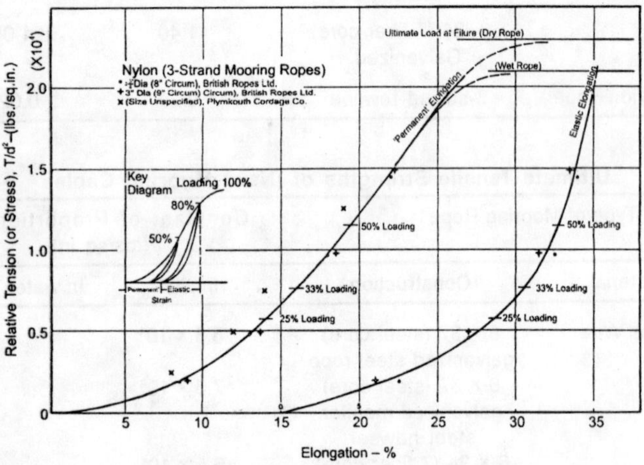

Fig. 5.7 : Generalized stress-strain for nylon mooring rope.

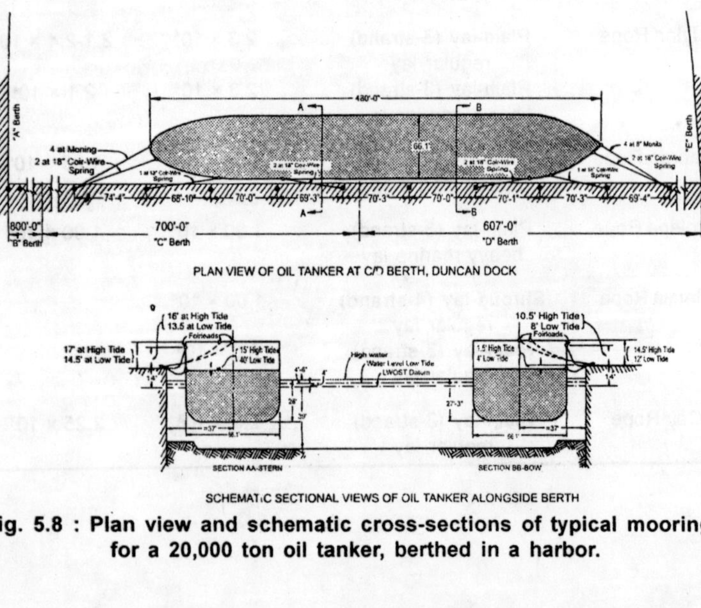

Fig. 5.8 : Plan view and schematic cross-sections of typical moorings for a 20,000 ton oil tanker, berthed in a harbor.

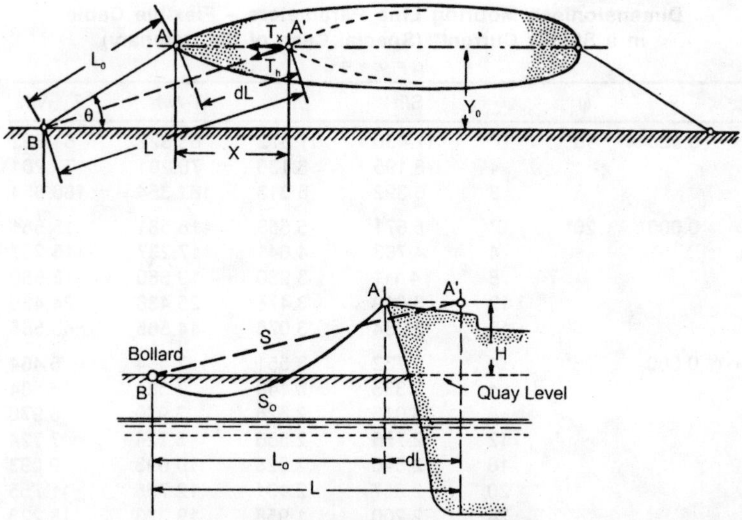

Fig. 5.9 : Mooring line geometry

tide level, variable cargo load and parasitic motions of surge, sway and heave, pitch, yaw and roll, the cumulative elastic behavior of the moorings poses a problem.

In analytical treatments of the surge of ships that have hitherto been made, some rather gross general assumptions have been made to render the problem tractable. Thus, the effect of all the ropes on surge motion has generally been regarded as concentrated in a net component of cumulative longitudinal tension T_X bearing some definitive relationship to the surge movement of the ship X measured from the equilibrium position A. The all-too-obvious fact that displacement of the ship from A to A′ inevitably pulls the bow inwards towards the quay and induces additional yaw and sway has been disregarded. Similar side motions result, of course, when the surge is forward and the pull is on the stern ropes. Full consideration of thes effects, as also of the heave, pitch and roll of the moored ship, would require an intricate analysis of coupled motions, which has not yet been attempted, except for *inelastic* mooring lines.

Despite these shortcomings, it is probable that the approximations yield reasonable estimates of the behavior of a moored ship under surge action.

Dimensionless Mooring Line Parameters – Flexible Cable in a Steady Current* (Special Case of the Catenary)

$\mu = \infty \; \gamma = 0$

μ	γ	ψ_n	ψ_{n-1}	S/h	l/h	T_n/wh	T_{n-1}/wh
∞	0.000	10*	0*	11.430	11.372	65.822	64.822
			4	8.195	8.130	78.201	77.201
			8	6.392	6.313	181.354	180.354
∞	0.000	20*	0*	5.671	5.553	16.581	15.581
			4	4.763	4.641	17.237	16.237
			8	4.111	3.980	19.580	18.580
			12	3.619	3.475	25.436	24.436
			16	3.234	3.075	44.565	43.565
∞	0.000	30*	0*	3.732	3.551	7.464	6.464
			4	3.333	3.149	7.584	6.584
			8	3.015	2.825	7.970	6.970
			12	2.756	2.556	8.724	7.724
			16	2.540	2.328	10.093	9.093
			20	2.357	2.131	12.755	11.755
			24	2.200	1.958	19.223	18.223
			28	2.062	1.804	52.174	51.174
∞	0.000	40*	0*	2.748	2.498	4.274	3.274
			4	2.539	2.288	4.309	3.309
			8	2.363	2.107	4.416	3.416
			12	2.213	1.950	4.611	3.611
			16	2.084	1.811	4.924	3.924
			20	1.970	1.685	5.411	4.411
			24	1.869	1.572	6.193	5.193
			28	1.779	1.467	7.552	6.552
			32	1.696	1.370	10.341	9.341
			36	1.623	1.278	18.825	17.825
∞	0.000	50*	0*	2.145	1.819	2.799	1.799
			4	2.028	1.700	2.812	1.812
			8	1.926	1.595	2.850	1.850
			12	1.836	1.499	2.916	1.916
			16	1.756	1.412	3.018	2.018
			20	1.684	1.331	3.165	2.165
			24	1.619	1.256	3.374	2.374
			28	1.560	1.185	3.676	2.676
			32	1.506	1.117	4.131	3.131
			36	1.455	1.053	4.866	3.866
			40	1.409	0.990	6.214	5.214
			44	1.366	0.929	9.395	8.395
			48	1.325	0.869	25.396	24.396
∞	0.000	60*	0*	1.732	1.317	2.000	1.000
			4	1.666	1.250	2.005	1.005
			8	1.607	1.189	2.020	1.020
			12	1.554	1.131	2.045	1.045
			16	1.506	1.077	2.084	1.084

contd....

contd....

μ	γ	ψ_n	ψ_{n-1}	S/h	l/h	T_n/wh	T_{n-1}/wh
			20	1.462	1.027	2.137	1.137
			24	1.421	0.978	2.209	1.209
			28	1.384	0.931	2.305	1.305
			32	1.349	0.886	2.436	1.436
			36	1.316	0.841	2.618	1.618
			40	1.286	0.798	2.879	1.879
			44	1.257	0.754	3.279	2.279
			48	1.229	0.711	3.955	2.955
			52	1.203	0.668	5.322	4.322
			56	1.179	0.623	9.444	8.444
∞	0.000	70*	0*	1.428	0.902	1.520	0.520
			4	1.393	0.867	1.521	0.521
			8	1.362	0.833	1.527	0.527
			12	1.333	0.802	1.537	0.537
			16	1.306	0.771	1.552	0.552
			20	1.282	0.742	1.572	0.572
			24	1.259	0.713	1.598	0.598
			28	1.237	0.684	1.632	0.632
			32	1.217	0.657	1.676	0.676
			36	1.197	0.629	1.732	0.732
			40	1.179	0.601	1.806	0.806
			44	1.162	0.573	1.906	0.906
			48	1.145	0.544	2.045	1.045
			52	1.129	0.515	2.249	1.249
			56	1.114	0.485	2.574	1.574
			60	1.099	0.453	3.163	2.163
			64	1.085	0.420	4.547	3.547
			68	1.071	0.383	11.487	10.487
∞	0.000	80*	0*	1.192	0.512	1.210	0.210
			4	1.177	0.497	1.210	0.210
			8	1.164	0.483	1.212	0.212
			12	1.152	0.470	1.215	0.215
			16	1.141	0.456	1.220	0.220
			20	1.130	0.443	1.226	0.226
			24	1.120	0.430	1.234	0.234
			28	1.111	0.416	1.244	0.244
			32	1.102	0.403	1.257	0.257
			36	1.093	0.390	1.273	0.273
			40	1.085	0.376	1.293	0.293
			44	1.077	0.361	1.318	0.318
			48	1.069	0.347	1.350	0.350
			52	1.062	0.331	1.392	0.392
			56	1.055	0.315	1.449	0.449
			60	1.048	0.298	1.531	0.531
			64	1.041	0.279	1.654	0.654
			68	1.035	0.259	1.862	0.862
			72	1.028	0.235	2.279	1.279
			76	1.022	0.209	3.536	2.536

This paper attempts to fill, in part, a long-vacant niche in the literature of mooring operations concerning the elastic qualities and behavior of mooring lines. While it probably cannot be said that there has been any great lack of knowledge on this score among long standing and reputable rope manufacturers, it can almost certainly be said that large users of rope, such as port authorities and shipping firms, have not always used their rope supplies to the greatest advantage and in the optimum fashion so as to eliminate needless failures and wastage. As ocean engineering operations move relentlessly towards deeper water, greater complexity and highter risk, the need for reliability in the mooring of objects, large and small, is ever more pressing. As the dynamic capabilities of moorigs of all types become of increasing importance, guidance from some basic principles for selecting ropes for specific tasks is greatly needed.

To this end, the weight and elastic properties of steel-wire and fiber mooring ropes are reviewed and generalized. Stress-strain relationships are shown to be separable into permanent-set and elastic parts, the latter of which can be described by a general empirical law governing all ropes. Under dynamic loading, the elastic modulus invoked is critically dependent on the magnitude of the initial tension in its relation to the ultimate strength of the rope. Here again it is possible to assign a general empirical law, governing this relationship, that is seemingly applicable to most rope materials.

Means whereby the elastic properties of anchor lines, suitable for deep or shallow water, may be construed in terms of their effects of restraint on a moored object are examined next. The general problem of a mooring line subject to current and wave action and the influences of the motions of the moored object is extremely complex and requires solution by digital computer procedures, but the essential part played by the leasticity of the anchor line is indicated. In general, the resulting effect on the moored object invokes a restraint-motion relationship that is nonlinear, odd, and nonsymmetric.

Mooring of ships in harbors involves mooring-line geometry which requires consideration of the fundamental properties of catenaries. The relationship between the elasticity of the mooring lines and the control of ship surging in the presence of water

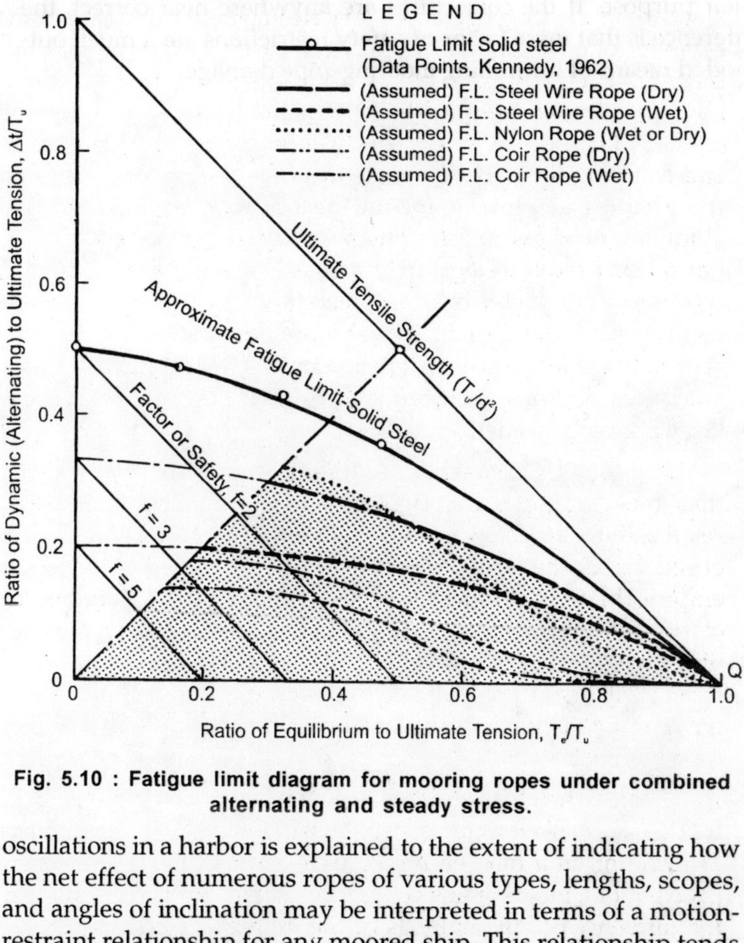

Fig. 5.10 : Fatigue limit diagram for mooring ropes under combined alternating and steady stress.

oscillations in a harbor is explained to the extent of indicating how the net effect of numerous ropes of various types, lengths, scopes, and angles of inclination may be interpreted in terms of a motion-restraint relationship for any moored ship. This relationship tends to follow a simple empirical law which is a nonlinear, odd, symmetric function.

The final part of this paper is devoted to a discussion of the fatigue properties of mooring ropes, which appears to be an area of relatively untrodden ground. By considering the known fatigue properties of solid steel, the probable endurance capacities of steel-wire and fiber ropes are surmised in terms of an alternating-stress, steady-stress diagram. Although results are highly conjectural, they may not be unreasonable. If they do no more than provoke researchers to explore this subject more fully, they will have served their purpose. If the conjectures are anywhere near correct, the inference is that mere factor-of-safety restrictions are a much outmoded means of controlling mooring-rope damage.

Chapter Six
The Nature of Waves in Oceans

6.1 Introduction

"Wind blowing over a water surface generates waves in the water by a process which connot be regarded as known". This statement is given by Ursell in 1956.

On the face of it, such a statement might appear extreme and altogether unwarranted, for the causal relation between wind and waves is apparent to all who have observed the sea, even if only with idle detachment. And, indeed, empirical relations which linked wind and sea and in which one could repose a great deal of confidence were already available at the time.

What was then lacking was an understanding of the physical mechanism whereby waves are generated and grow under the spurring action of the wind. Twelve years ago, the three elements essential to such an understanding, namely, a reliable fluid dynamic theory, a comprehensive set of accurately executed laboratory tests and field trials, and a satisfactory correlation between theory and measurements, were all wanting. The picture as to how waves were generated was incoherent. The difficulties that stood in the way of a formulation were not mathematical but, more importantly, conceptual. Today, the situation is much improved. Although a considerable amount of development remains to be brought to a successful conclusion, there is now a working hypothesis.

This change has been brought about chiefly by two investigators: Phillips and Miles. To place their work in the proper light, it is best to begin with a brief exposition of the development that preceded their efforts.

Now let us try to understand about gravity waves. The first step toward an explanation of how gravity waves are generated by wind action appears to have been taken a century ago by Helmholtz, who suggested a mechanism based on surface instability. Such a suggestion was an obvious first approach, for a stability analysis is relatively simple since the excitation is not introduced in the balance of forces.

His primitive mathematical model was based on the coupling between surface waves (deterministically described) and coherent atmospheric fluctuations which are wave-induced. The balance of forces accounts only for normal pressures and gravitational restoration. Instability is considered to result when the suction of the wind on a wave crest exceeds the gravitational restoring reaction. Since the mathematical model rests on a neglect of the important roles played by viscosity and surface tension, the conclusion Helmholtz derived, namely, that the slightest breath of wind would ruffle the surface of the sea, is at odds with observations.

Kelvin expanded the Helmholtz model (Thomson, 1871) to include surface tension and came to the conclusion that the calm surface of the water would be unstable if the wind exceeded 646 cm sec^{-1} or, roughly, 12.5 knots. This minimal velocity is not in concord with observations. Although there is a great deal of controversy as to whether there does, indeed, exist a critical wind speed for wave generation, such a speed (if it does exist) would appear to be considerably lower. Kelvin himself realized that his hypothesis did not reflect reality and suggested that viscosity might play a role, thus anticipating a hypothesis of Jeffreys.

It should be noted that the instability mechanisms of Helmholtz and Kelvin do not really explain how waves in water are initiated. The mechanisms only explain whether a perturbation of the water surface having the characteristics of a wave will tend either to grow or to decay under the action of the wind once it has been developed by an unspecified external cause other than the wind itself. Thus, an instability mechanism provides, at best, only a partial answer.

A lesson to be derived from the models of Helmholtz and Kelvin relates to the importance of accounting for all the reactive forces called into play: inertia, gravity, surface tension, and viscosity. The omission of any one of these might well lead to unrealistic conclusions.

The influence of viscosity was considered in the same physical model by Wuest and Lock. Since the former is a simplified approximation to the latter, it is sufficient to consider only the second one.

When air flows over the surface of water, boundary layers are formed in both media through the action of viscosity. Within the boundary layers, the wind intensity varies with height while the water is in motion; beyond the boundary layers, the wind intensity is constant and the water is calm. Under conditions of low wind intensity and short fetch length, the boundary layers are, at least initially, laminar. These conditions are assumed to hold.

The first step is to calculate the motions in the boundary layers for a rigid interface, and the second is to examine whether the motions remain stable when perturbed by an infinitesimal oscillation.

If the interface is rigid and the frequency of the oscillation controllable, the perturbation in the flow (either air or water) decays at first. Farther downstream, it may pass through a region where it is amplified. If this amplification is sufficient, the flow becomes turbulent. If the amplification is insufficient to bring about transition to turbulent flow, the perturbation decays farther downstream.

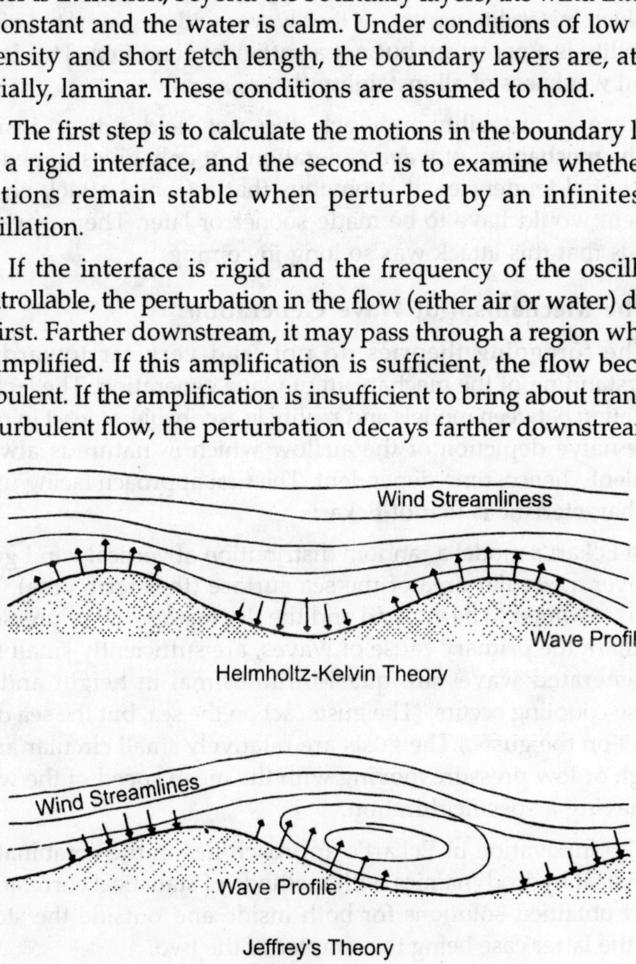

Fig. 6.1 : Wave Generation models.

When the interface is no longer rigid but is allowed to take form in response to the governing forces, the problem increases in complexity, but the general behavior derived for a rigid interface continues to obtain.

The model indicates that surface instability does occur for certain frequencies even with relatively low wind speed (100 cm sec^{-1} or 0.2 knots). Of course, this does not imply that the waves generated will be of frequencies corresponding to the spectral range for which instability occurs, or that waves will be generated at all! Instability is a necessary but not a sufficient condition. This is the central weakness of all instability theories.

Since an instability approach does not yield a direct insight into the mechanism of wave generation but only gives some idea as to initial tendencies, it is obvious that a frontal attack on the problem would have to be made sooner or later. The surprising thing is that this attack was so long in coming.

6.2 The Mechanism of Wave Generation

The foregoing theories do not lead very far toward an understanding of the mechanism of wave generation. The lack of correlation between models and reality is ascribable, at least in part, to the naive depiction of the airflow which in nature is always turbulent - hence, time-dependent. The first approach facing up to this characteristic is that of Eckart.

In Eckart's model a random distribution of vertical wind gusts acts over a circular area of the sea surface (the storm area). The fluid is assumed to be inviscid and the fluctuations of air pressure, which are the primary cause of waves, are sufficiently small that the generated waves are quasi-infinitesimal in height and no reverse coupling occurs. (The gusts 'act on the sea, but the sea does not act on the gusts.) The gusts are relatively small circular areas of high or low pressure moving with the mean speed of the wind and having a specific duration.

The innovation of Eckart's approach lies in the combination of classical hydrodynamics with methods of statistical correlation. Eckart obtained solutions for both inside and outside the storm area, the latter case being the simpler of the two.

Inside the storm area, each point is traversed by waves having a spectral distribution of energy and heading in a wide range of directions. This makes for a confused sea. The confusion is greatest at the center of the storm area. As one proceeds toward the periphery of the storm, the confused area begins to manifest a predominant direction of wave propagation which becomes stronger as the distance from the storm center increases. Thus, the wave spectrum is a function of its position within the storm area.

Outside the storm area, the situation is simpler because of the absence of excitation. The direction of wave propagation tends to be radial at great distances from the storm center, and the root-mean-square wave height tends to vary as the inverse square root of the radius vector.

The picture is correct but only in a qualitative sense, for the method fails in the prediction of wave height by an order of magnitude. A reason can be advanced for this want of quantitative correlation between model and reality. The storm is defined as an ensemble of traveling pressure areas and the kinetic influence of wind velocity on wave formation (e.g., Jeffreys' sheltering) is neglected. This neglect is consistent with the assumption of infinitesimal wave height; but as the waves tend to grow, mechanisms of kinetic energy transfer from wind to water, which depend on the wave slope and higher powers thereof, make their appearance and perhaps eventually become dominant.

Radical Change in the State of the Art

This was the state of the art at the time of Ursell's observation. It can be summarized as rather unsatisfactory. But in 1957 the picture radically changed with the introduction of two new hypotheses which are complementary to each other : the resonance hypothesis of Phillips and the shear-flow hypothesis of Miles.

6.3 Phillip's Theory and Mile's Shear-Flow Hypothesis

In Phillip's theory," it is assumed that there is no return coupling between surface waves and wind (i.e., the excitation does not depend on the response). This is the same assumption made by Eckart, but Phillips introduced a much more realistic representation of the wind and assumed that the dimensions and lifetimes of the gusts are describable by a spectrum, i.e., by a quasi-infinite set of sinusoidal pressure fluctuations of varying amplitude superposed in random phase. Of course, the generated waves are also describable by a spectrum.

The central result is that the frequency-response function contains a selector for wave growth by resonance. When the spectral component of surface pressure moves at a relative convective velocity $U_c \cos(\theta)$ equal to the celerity of the free surface wave of the same number, the latter tends to grow linearly with time. In the expression, θ is the angle between the convection velocity and the wave heading.

The waves develop in two stages. In the initial stage, which lasts until the pressure fluctuations are fully developed, the generated waves have a spectral peak at I. 7cm and a phase velocity.

$$C = (4g\,T/\rho)^{1/4}$$

where T is the surface tension. They move in the directions $\cos^{-1}(C/U_C)$ to the mean wind. Observations by Roll lend qualitative support to this finding.

The greatest amount of wave growth occurs during the second stage, which continues until the waves peak so high that nonlinearities become important. The wave spectrum is related to the pressure spectrum in such manner that the following relation obtains:

$$\overline{\eta^2} = \overline{p^2} \cdot \frac{t}{2[2\rho^2 U_c g]^{1/2}}$$

where $\overline{\eta^2}$ and $\overline{p^2}$ are, respectively, the mean-square surface displacement and pressure fluctuation and t is the elapsed time. Note that, in this model, the growth of surface-wave energy (proportional to $\overline{\eta^2}$) is linear with time.

The result explains at once the inability of Eckart's theory to predict correctly the wave height. His imprecise definition of pressure distribution obliterates the resonance peak of the response of the water surface. It is precisely the wave components near the spectral peak that account for the major contribution to the wave spectrum.

To predict wave height, the mean pressure fluctuation $\overline{p^2}$ must be known. Phillips obtained excellent agreement with observational data on $\overline{\eta^2}$ by overestimating $\overline{p^2}$ by an order of magnitude. This tends to cast doubt as to the validity of Phillips' resonance mechanism of wave generation; however, a conclusion is premature. Before arguing it further, it is necessary to consider the model of Miles.

Miles followed the lead of Helmholtz, Kelvin, and Jeffreys and considered that energy is transferred from wind to waves by pressure distributions related to wave form. The model, therefore, is a coupled one. Again the assumption is made that the water is an ideal (inviscid) fluid. This leads to the application of the convenient methods of potential theory. As to the air, turbulence is introduced, but only insofar as it determines the average wind profile U(z). Thus, in the original model, the airflow (except for its profile) is laminar. This is, of course, an artificiality. It is precisely this artificiality that gives an unexpected result when a determination of instability is made.

According to Miles' mechanism, within a zone termed the *critical layer*, energy is first transferred from the mean wind to the wave-coupled oscillations in the airstream. Then the energy is again transferred as Reynolds stresses to the water waves, the rate of energy transfer being given by

$$dE/dt = \rho_a \int_0^\infty \overline{uw}\ [\delta U(z)/\delta z] dz$$

where u and w are, respectively, the horizontal and vertical components of the wind speed U(z). The bar indicates that their temporal mean is to be taken. The product $\rho_a \overline{uw}$ represents the Reynolds stresses. These obtain only in the critical layer and are zero for $z = z_c$ where z_c is the depth of the cirtical layer.

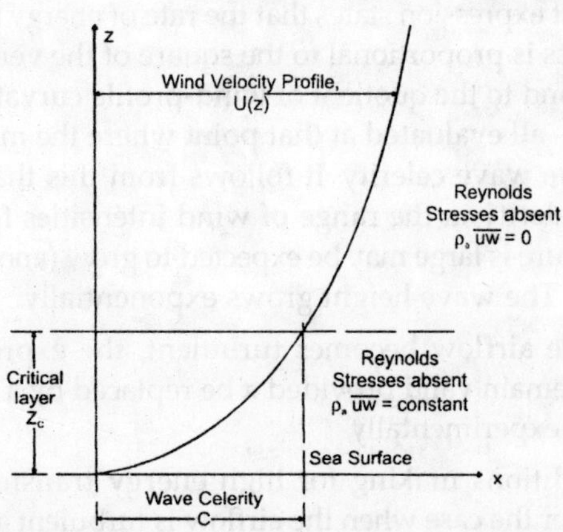

Fig. 6.2 : The critical layer according to Miles.

The critical layer is the layer of fluid for which the mean speed equals the celerity of the wave perturbation.

$$U(z_c) = \frac{1}{z_c}\int_0^{z_c} U(z)dz = C$$

For numerical prediction it is necessary to introduce a mean velocity profile. Miles used the logarithmic law, where k is von Karman's universal constant for turbulence (k = 0.4) and z_o is an effective roughness parameter. The validity of such a law is questionable when applied to flow over water waves.

$$U(z) = (U^*/k) \bullet ln(z/z_o)$$

The central results of Miles' theory are contained in two expressions. The flux of momentum from wind to waves is given by

$$\tau = \rho_a \overline{uw}$$

$$= \frac{\pi\rho_a}{k}\left\{\overline{w^2(z)} \bullet \frac{d^2U(z)/dz^2}{dU(z)/dz}\right\}_{Z=Z_c}$$

while the coupling parameter is

$$\mu = \frac{2\tau}{\rho C^2 k^2 a^2(k)}$$

Here $\overline{w^2}(z_C)$ is the mean square of the vertical induced-velocity fluctuations within the critical layer. The first expression is evaluated at the depth of the critical layer; a is the amplitude of the wave component of interest. Since the rate of energy transfer is given by τC, the first expression states that the rate of energy transfer from wind to waves is proportional to the square of the vertical velocity fluctuations and to the quotient of wind-profile curvature to wind-profile slope - all evaluated at that point where the mean airspeed is equal to the wave celerity. It follows from this that only those waves of a celerity in the range of wind intensities for which the profile curvature is large may be expected to grow (another selective mechanism). The wave height grows exponentially.

When the airflow becomes turbulent, the expressions (and conclusion) remain valid provided π be replaced by a coefficient to be evaluated experimentally.

The conditions making for high energy transfer have been determined for the case when the airflow is turbulent and the water surface is aerodynamically smooth, in which case Miles' mechanism appears to be most effective for wave components whose phase velocity lies between

10 U* and 20 U*

where U* is the shear stress velocity defined by

$$U^* = (\tau_0/\rho_a)^{1/2}$$

and τ_0 is the surface stress.

It is now possible to discuss the resonance model of Phillips and the shear-flow model of Miles. Recall that in Phillips' model the mechanism of wave generation is the direct action of the turbulent pressure fluctuations on the sea surface and that the pressure field is uninfluenced by the wave pattern. In Miles' model, turbulence itself is neglected (except that the assumed mean wind profile must be that occurring in turbulent flow). The mechanism is one of interaction between mean flow and waves. The two mechanisms are not mutually exclusive. Indeed, the hypothesis can be introduced that they act concurrently. Miles took this step.

Since both the resonance and shear-flow model are linear, they can readily be combined by parametric proportioning of the component models. It is possible to take as much of one or of the other as observational evidence suggests.

The combined model indicates that the growth of a wave component is at first linear. After some time it passes gradually to exponential. This is as far as the linear combined model goes. Eventually, nonlinear phenomena and other (dissipative) processes not considered in the model make their presence felt, and a saturation height is attained.

The time a generic wave component requires to complete the stage of linear growth is determinable as a function of C/U and relative wave direction θ. This time grows quasi-exponentially with C/U, amounting to about 100 wave periods for C/U = 0.2, to close to 1000 wave periods for C/U = 0.5, to close to slightly over 10,000 wave periods for C/U close to unity. This implies that the growth of all waves is initiated by the resonance mechanism and that this tends to be the only mechanism which contributes to the growth of wave components of low C/U ratio. At high C/U ratios, the growth is determined essentially by the mechanism of shear flow.

Confirmatory data are difficult to obtain in the laboratory and even more difficult at sea. But the few sets available tend to support the combined model.[2,8,11,12,18] The comments by Longuet-Higgins are particularly remarkable in this connection.

He indicates that the shear-flow model gives better predictions than does the resonance model of the angular spread of the wave spectra and of the shape thereof (the resonance model predicts a double-peaked distribution of energy which is not observed). In addition, in the region of the energy peak, the correlation between wind-pressure fluctuations and wave motion favors the shear-flow model.

6.4 The Probability Distributions

Probability distributions of the extreme values of a variate have been studied by various mathematicians, notably Gumbel, who made the first compilation of the techniques. The kernel of Gumbel's work is a set of three asymptotic distributions. Of these, the first (sometimes referred to as a Fisher-Tippett Type I distribution) has been extensively used in meteorological predictions.

Davenport has made a plea for its adoption as a standard method of analysis in Canada. Bell has used it to analyze wind records in Hong Kong. Whittingham has used it in the study of extreme winds in Australia. Court analyzed wind speeds at 25 localities in the United States and has found good agreement between theory and observation. These are but a few references. The first asymptotic distribution has been also applied to waves and ship motions and stress intensities with excellent results.

The first asymptotic distribution is not bounded. This means that the expected maximum value will tend to grow without limit as the interval of prediction approaches infinity. It is fairly obvious that if this distribution has any bias, it is in the direction of overestimate. The predictions must be viewed as conservative. The objection against an unlimited expected value is, perhaps, serious only in principle, for the variate approaches infinity with logarithmic slowness, i.e., at a very lethargic pace.

The inputs required for an estimate of expected maximum values are a tabulation of maximum attained values over a specific interval, usually a year. To estimate the maximum expected values in the very long run, the maximum attained values are arranged in descending order, the index of ranking being denoted by i. Each attained maximum value x is assigned a plotting position given by the frequency value $p_i = (N-i)/N$, where N is the number of intervals of observation. The mean of the observed values $\overline{x}_N$ and the corresponding standard deviation σ_N and variance σ^2_N are calculated from

$$\overline{x}_N = \frac{1}{N} \sum_{i=1}^{N} x_i$$

$$\sigma^2_N = \frac{1}{N} \sum_{i=1}^{N} (x_i - \overline{x}_N)^2$$

$$\sigma_N = \sqrt{\sigma^2_N}$$

A reduced variate is introduced and defined as

$$y = a\,[x - u]$$

where a is a slope and u is a mode. This variate is related to the frequency value by

$$y = -ln(-\ln(p))$$

The slope is determined from

$$a = E(\sigma_N)/\sigma_N$$

where $E(\sigma_N)$ is the expected value of the dispersion of N values (while σ_N is the actual value).

The mode is determined from

$$u = \overline{x}_N - E(y_N)/a$$

where $E(y_N)$ is the expected value of the reduced variate.

Both expectd values are funtions of the total number of observations. The differences, however, between the expected values for an infinite number of observatios and for resonably few (25 to 50) is quite small so that no appreciable error is introduced by using the values corresponding to infinite observations, in which case:

$$E(\sigma) = E(\sigma)_N = \pi/\sqrt{6}$$

$$E(y) = E(y_N) = 0.5772 \text{ (Euler's number)}$$

The curve of extremes is, then,

$$x = u + y/a$$

$$= \overline{x}_N + \sigma_N \sqrt{6/\pi}\,[y - 0.5772]$$

6.5 An Explosion Wave System

An explosion wave system is produced by the tendency of the free surface to return to the mean level under gravitational forces, following the deformations and velocities imparted to it by the explosion.

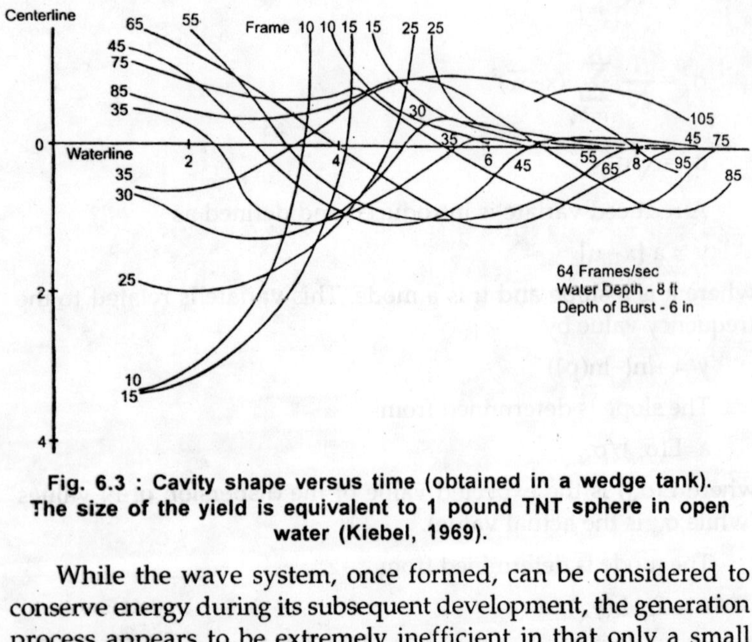

Fig. 6.3 : Cavity shape versus time (obtained in a wedge tank). The size of the yield is equivalent to 1 pound TNT sphere in open water (Kiebel, 1969).

While the wave system, once formed, can be considered to conserve energy during its subsequent development, the generation process appears to be extremely inefficient in that only a small fraction of the total available energy is radiated as organized wave motion. As much as 50% of the thermal energy available in a submerged (nuclear) explosion is lost as irreversible heating or shock in the water during the first bubble expansion. Much of the remainder (excepting that small fraction appearing as wave motion) generates disorganized turbulent motion and then dissipates to heat. This phenomenon is readily apparent for an explosion at shallow depth by the successive appearances of the massive spray dome, column, plume and base surge as precursers to the issuance of waves from the central region. For detonations at great depth or above the free surface, the losses increase with the result that significant waves occur for only a restricted range of depth of burst. In the case of deep submergence (see Cole, 1948; Kot, 1964, among others) the explosion bubble undergoes repeated oscillations which radiate successive shock waves carrying away energy. (The bubble may, in fact, collapse entirely, leaving a mass of warm fluid and essentially no surface disturbance.) A detonation above the surface is preferentially dissipated into the atmosphere; the waves generated by the surface overpressure are consequently small.

It may be envisioned, then, that the significant generation mechanism is the venting of the bubble at which instant the free surface is effectively "cratered." This initial crater, with dimensions determined by yield and depth of burst, collapses under gravity and produces the outward radiating wave train.

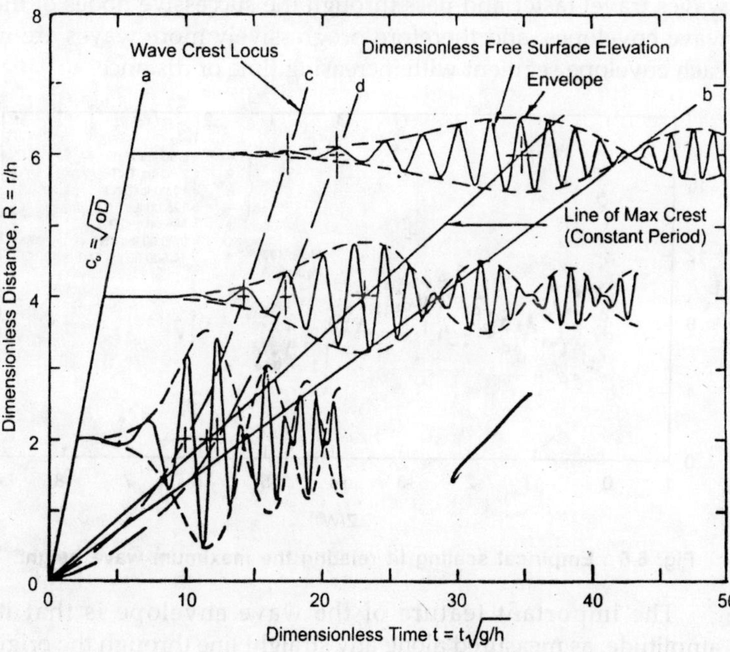

Fig. 6.4 : Theoretically calculated wave trains.

The subsequent propagtion shows three successive stages in the development of the wave train. The vertical and horizontal axes in this figure are dimensionless distance R = r/h and dimensionless time T = $t\sqrt{g/h}$, respectively. The three curves show the amplitude-time histories of a wave train generated at the origin as they would be recorded at nondimensional distances, R, of 2, 4 and 6. The symmetrical dashed curves bounding the wave train comprise the wave envelope and serve to define the distribution of energy within the train. The precise shape of the envelope depends on the initial source conditions, whereas the space-time coordinates of the individual waves are independent of the source and depend only on the water depth. A characteristic of the wave envelope is that any identifiable portion of it-say, a node or antinode-propagates at uniform velocity, as shown by the straight lines a and b. On the other hand, the trajectory of a particular wave is a curve, concave upwards, because the waves are continuously accelerating towards the limiting phase velocity C = $\sqrt{gh}$ of the wave front. Thus, the waves travel faster and pass through the successive nodes of the wave envelopes, and therefore progressively more waves are in each envelope segment with increasing time or distance.

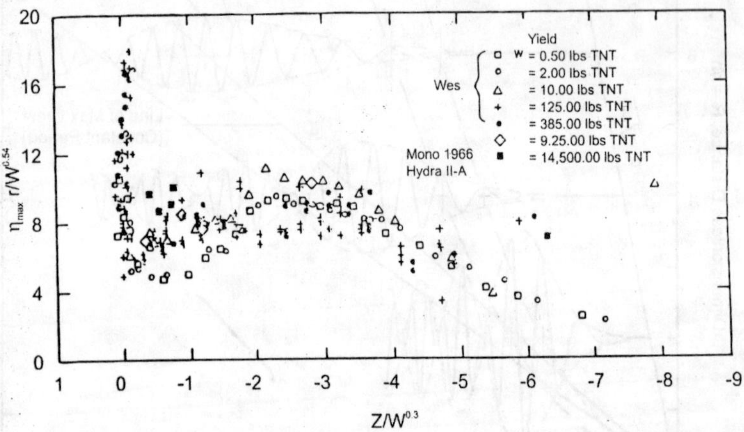

Fig. 6.5 : Empirical scaling fit relating the maximum wave height.

The important feature of the wave envelope is that its amplitude, as measured along any straight line through the origin of the R - T diagram, is inversely proportional to its distance from the origin, owing to the combined effects of circumferential spreading and the stretching of individual groups. Thus, the height of the highest wave in the upper wave train is about 1/3 that of the corresponding wave in the lower train, the former having traveled three times farther.

6.6 Propagation of Waves in Nonuniform Depth

As the expanding cirular wave front encounters depth variations it suffers the usual effects of refraction and shoaling. Accounting for these is complicated by the circular and dispersive character of the wave system. For example, wave elements of different frequencies initially travelling in the same direction will gradually diverge owing to their differing rates of refraction. The wave history at any point is then the superposition of components coming from different directions. The shoaling problem is correspondingly difficult.

A useful approximate procedure has been developed by Van Dorn (1964). (See also Van Dorn, LeMéhauté and Hwang, 1968, or LeMéhauté, 1971. An equivalent development has been given by Van Mater and Neal, 1970.) This method requires that bottom contours be approximated by straight lines (not necessarily parallel) and that the laws of geometirc optics apply. The latter requires; 1) that the wave amplitude be small; 2) that the water depth vary slowly over a wave length; 3) that the phase velocity at any point be that of the corresponding constant depth.

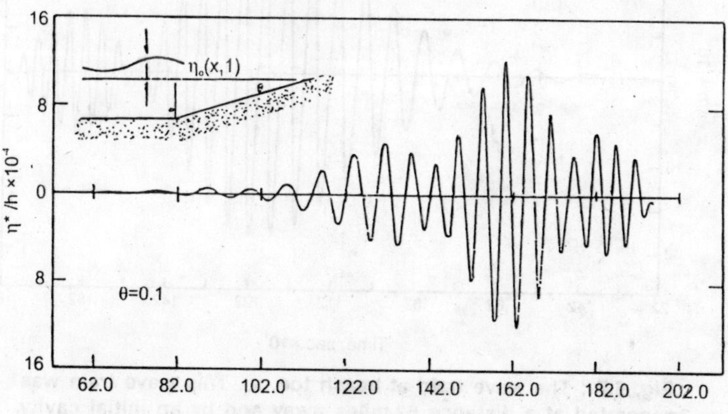

Fig. 6.6 : Theoretical wave profile showing the interference between incident and reflected wave trains at $x_0/4$ from the shore.

The foregoing scheme is practical only for digital computer application; the appropriate equations, therefore, need not be reproduced here but can be found in the cited sources. For many conditions of practical interest, the wave front propagates a sufficient distance before entering shoaling water that it may be considered effectively one-dimensional. In such a case, and if the angle of incidence is not too oblique, a crude estimate of the wave characteristics can be obtained by simply applying refraction and shoaling coefficients to the amplitude envelope. The justification for this is two fold. First, after a long propagation distance the wave envelope tends to contain a great number of waves with very slow variation in height and frequency; in a sense, they are so well dispersed as to be locally periodic. Second, the rate of frequency dispersion lessens in shoaling water as the phase and group velocities approach equality. Under these conditions the system is adequately treated as a succession of periodic waves with the attendant simplifications. Since fixed points on the amplitude envelope are associated with fixed values of frequency, shoaling and refraction corrections are straightforward.

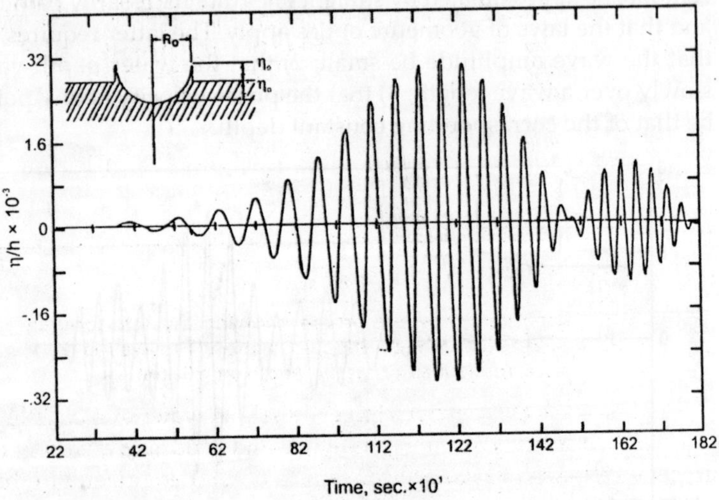

Fig. 6.7 : The wave train at beach toe, x_0. This wave train was generated at a distance 62 miles away and by an initial cavity, $_0$=300 feet deep and R_0 = 3300 feet in radius. The water depth is 16,400 feet. (Hwang et. al., 1969)

An additional possibility is the occurrence of reflection from depth variations (excluding at this point shreline reflection). The many theoretical attempts to derive a suitable formulation for the reflection coefficient for periodic progressive waves in water of nonuniform depth have been reviewed by LeMéhauté (1967). Of these, the works of Miche (1944) and Roseau (1952) have the most general application to the problem of explosion waves advancing shoreward from deep water. The Roseau theory applies to a particular family of bottom profiles, some of which quite closely resemble the continental shelves and their terminal escarpments. Application of this theory to the explosion wave problem (LeMéhauté, Whalin et al., 1966) indicates that in practical cases correction for wave reflection is insignificant for all frequencies of interest, even the long leading waves. In the limiting case of small reflection, the Roseau theory tends to the linear conservation of energy flux (in one dimension). This conclusion is further supported by wave tank experiments (LeMéhauté, Snow and Webb, 1966).

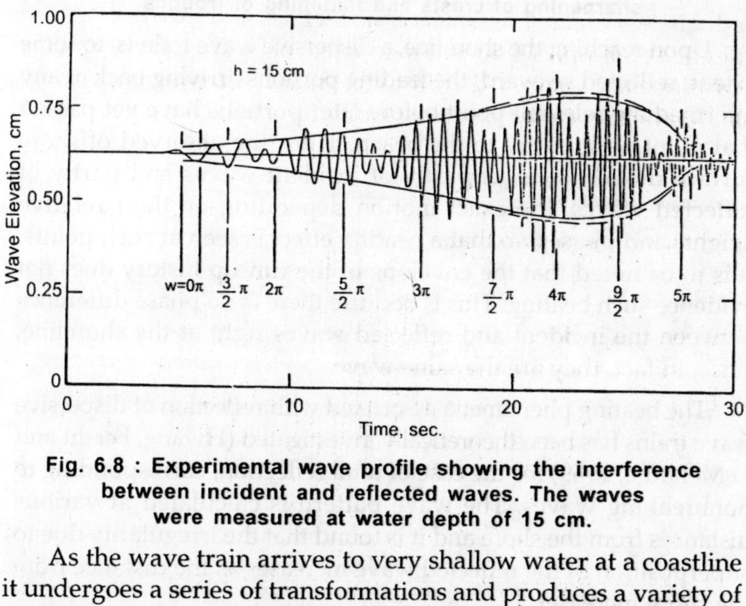

Fig. 6.8 : Experimental wave profile showing the interference between incident and reflected waves. The waves were measured at water depth of 15 cm.

As the wave train arrives to very shallow water at a coastline it undergoes a series of transformations and produces a variety of effects which are of primary concern to coastal engineers. In this section, we consider first nonbreaking wave effects including beating, run-up, nonlinear phenomena and harbor oscillations. Subsequently, attention is given to breaking waves and the development of wave set-up and shelf oscillations.

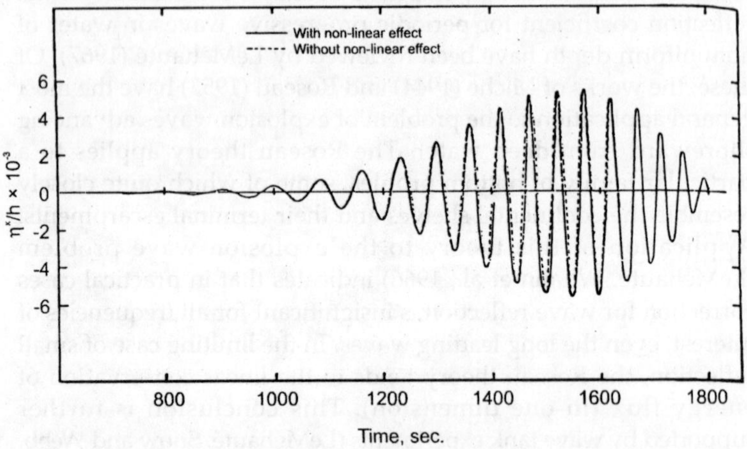

Fig. 6.9 : Wave run-up and the influence of nonlinear effects showing sharpening of crests and flattening of troughs.

Upon reaching the shoreline, a dispersive wave train is, to some extent, reflected seaward, the leading portions arriving back at any intermediate relevant point before later portions have yet passed that point on the way to the beach. Thus, the observed offshore surface motion consists partly of incident waves and partly of reflected waves, the exact motion depending on their relative heights and phases, so that a beating effect is seen at such points. It is to be noted that the envelope of the run-up history does not evidence such beating. This is because there is no phase difference between the incident and reflected waves right at the shoreline, since, in fact, they are the same wave.

The beating phenomena associated with reflection of dispersive wave trains has been theoretically investigated (Hwang, Fersht and LeMéhauté, 1969) for the case of total reflection, corresponding to nonbreaking waves. The wave pattern is calculated at various distances from the shore and it is found that the irregularity due to superposition of the reflected wave increases as the distance from the shore increases.

As the wave train enters very shallow water, the deep water equations become progressively inadequate. A more suitable formulation has been developed by Hwang, Fersht and LeMéhauté (1969) which is specifically adaptable to explosion wave trains. The primary feature not seen from a linear analysis is a steepening of crests and a flattening of troughs, in accordance with finite amplitude periodic wave theories.

The dashed and solid curves represent the linear and nonlinear formulations, respectively. The crest steepening is apparent. There is, however, no difference in peak displacements so that, for most applications, the linear development of Hwang et al. will suffice. It may be noted here that even in the limit of zero depth (i.e., the run-up history) the peak excursions are as given by the linear analysis, only the intermediate time history being influenced by the higher order terms.

The climb of a wave on an impermeable beach has received extensive investigation which will not be detailed here. Many of the relations determined for periodic waves may be applied to the explosion wave system if a judicious choice of parameters is made; the run-up summary of LeMéhauté, Koh and Hwang (1968) will be most useful in this regard.

A method was developed specifically for dispersive wave trains by Hwang, Fersht and LeMéhauté (1969) and has been mentioned regarding nonlinear effects. Basically, their analysis proceeds from a transformation of variables (Carrier and Greenspan, 1958) which maps the moving shoreline into a fixed line, thereby making run-up determination quite simple.

We turn now to a particularly important aspect of coastal explosion wave effects. An explosion train contains significant waves with a broad range of frequencies, the early arrivals being very long and the later being progressively shorter. Therefore, if a train is incident at a given harbor, it will at some time during its history supply an excitation at the fundamental harbor mode and at all harmonics. If the exciting waves are of even relatively small height, the harbor agitation may grow to significant levels. The same argument applies to ship mooring characteristics with a consequent threat of broken lines, damaged fenders and so on.

The problem is aggravated by the fact that explosion waves from a sizeable source tend to show the greatest amplitude for frequencies lower than ordinary sea or swell. The harbor, designed for sea and swell protection, may be particularly sensitive to this lower frequency excitation.

In order to estimate harbor effects one must first determine the harbor's response function. This can be done by using a method recently developed by Hwang and Tuck (1970). Their procedure gives the amplification factor versus incident wave number and was specifically developed for harbors of arbitrary shape. Furthermore, no entrance approximation is involved (such as the arbitrary imposition of a nodal line); instead, the harbor and the sea are considered as a whole, not separately. The response curve assumes periodic excitation, but it can be applied to an explosion wave train if there are a sufficient number of waves in each envelope that the wave to wave frequency variation is small.

6.7 The Breaking of Wave Transformation

Once broken the wave dissipates energy and diminishes in height as it proceeds shoreward.

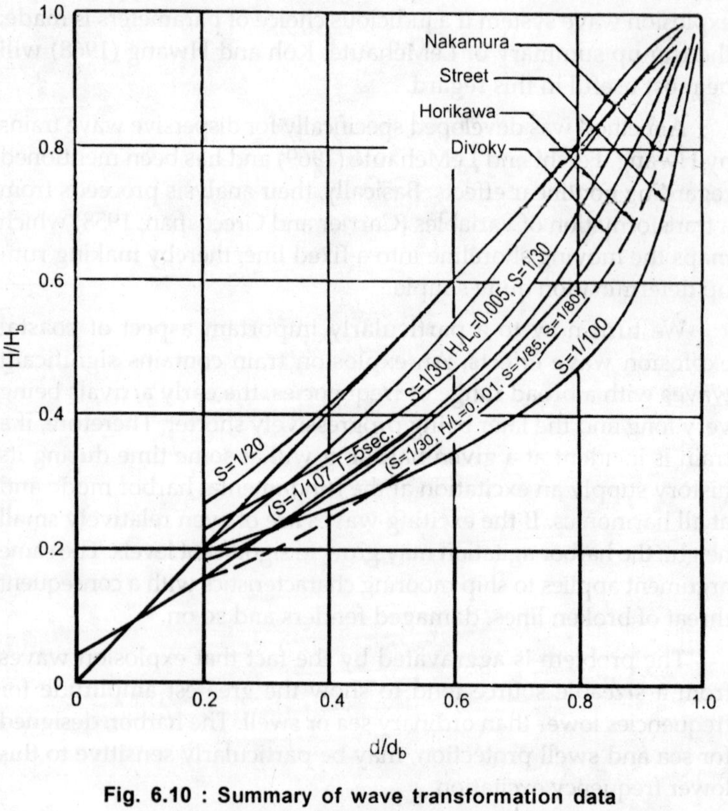

Fig. 6.10 : Summary of wave transformation data (shoreward of the breaking point).

The processes of shoaling, breaking dissipation, bottom friction dissipation and convergence or divergence of orthogonals through refraction are all operative in a real environment and contribute to a great complexity of surf-zone behavior. We shall consider the idealized case of normal incidence at a plane beach.

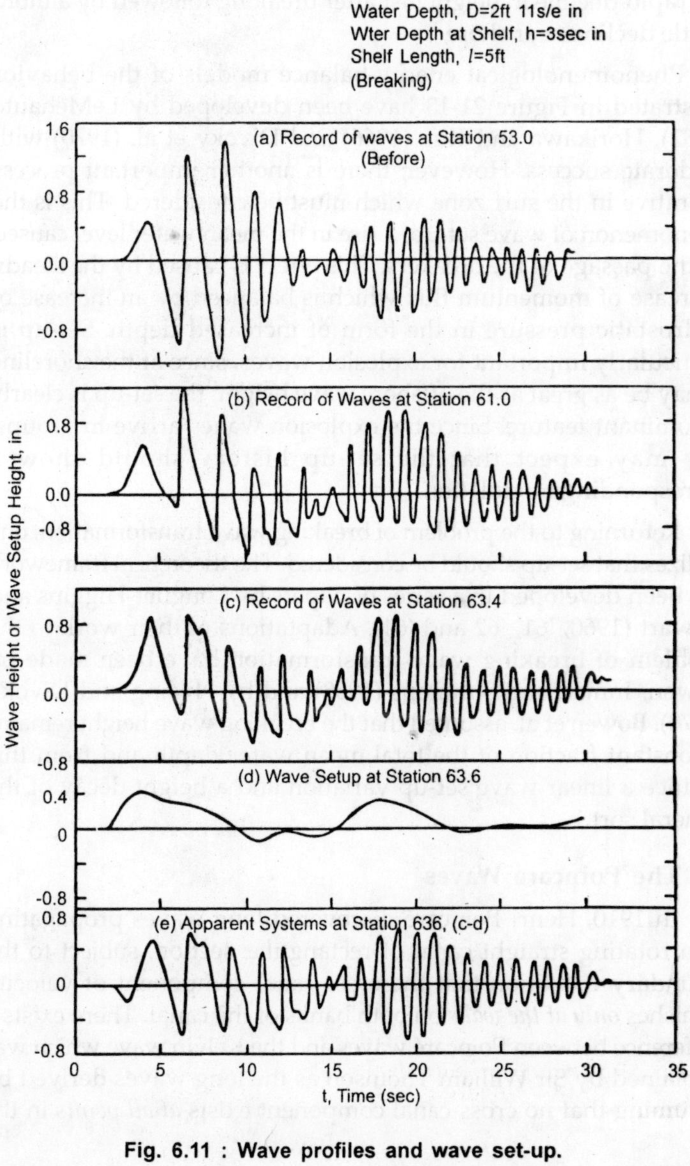

Fig. 6.11 : Wave profiles and wave set-up.

Experiments on wave height transformation have been reported by Street and Camfield (1966), Horikawa and Kuo (1966), Nakamura et al. (1966), and Divoky et al. (1970) among others. A summary of experiments is shown in Figure 21-13. Typically there is a rapid decline in height just after breaking followed by a more gentle decline near shore.

Phenomenological energy-balance models of the behavior illustrated in Figure 21-13 have been developed by LeMéhauté (1962), Horikawa and Kuo (1966) and Divoky et al. (1970) with moderate success. However, there is another important process operative in the surf zone which must be considered. This is the phenomenon of wave set-up, a rise in the mean water level caused by the passage of the breakers. This effect is caused by the steady decrease of momentum flux which is balanced by an increase of hydrostatic pressure in the form of increased depth. Set-up is particularly important for explosion waves, since at the shoreline it may be as great as the offshore wave height; the set-up is clearly a dominant feature. Since the explosion waves arrive in groups, one may expect that the set-up history should show a corresponding modulation.

Returning to the problem of breaking wave transformation, one realizes that set-up should be considered. The theoretical framework has been developed in a series of papers by Longuet-Higgins and Stewart (1960, '61, '62 and '64). Adaptations of their work to the problem of breaking wave transformation have been made by Bowen, Inman and Simmons (1968) and by Hwang and Divoky (1970). Bowen et al. assumed that the breaking wave height remains a constant fraction of the total mean water depth and from this deduce a linear wave set-up variation and a height decay of the general sort.

6.8 The Poincare Waves

In 1910, Henri Poincaré discovered long waves propagating in a rotating straight canal of rectangular section, subject to the boundary condition that the cross-canal component of velocity vanishes *only at the walls* on both banks of the canal. There exists a difference between Poincaré waves and the Kelvin wave which was explained by Sir William Thomson as the long waves derived by assuming that no cross-canal component exists *at all points* in the

Coefficients of $\sin \frac{n\pi}{2}\left(1-\frac{y}{a}\right)$ and $\cos \frac{n\pi}{2}\left(1-\frac{y}{a}\right)$ in the Function ζ_0 (y), U_o (y) and υ_o (y)

$\left(\frac{\pi a}{2\lambda}\right)$	β	n	$\dfrac{\left(\frac{\pi a}{2\lambda}\right)}{\frac{1}{\sqrt{2}}=\left\{\left(\frac{\pi a}{2\lambda}\right)^2+\left(\frac{\pi a}{2}\right)^2 f^2\right\}^{1/2}}$ (Unit : A)	$\dfrac{\left(\frac{\pi a}{2\lambda}\right)\cdot f}{\frac{1}{\sqrt{2}}=\left\{\left(\frac{\pi a}{2\lambda}\right)^2+\left(\frac{\pi a}{2}\right)^2 f^2\right\}^{1/2}}$ (Unit : A)	$\dfrac{\beta f}{\frac{1}{\sqrt{2}}=\left\{\left(\frac{\pi a}{2\lambda}\right)^2+\left(\frac{\pi a}{2}\right)^2 f^2\right\}^{1/2}}$ unit : (g/2ωa) • A	$\dfrac{\left(\frac{\pi a}{2\lambda}\right)\cdot\frac{n\pi}{2}}{\frac{1}{\sqrt{2}}=\left\{\left(\frac{\pi a}{2\lambda}\right)^2+\left(\frac{\pi a}{2}\right)^2 f^2\right\}^{1/2}}$ unit : (g/2ωa) • A	$\dfrac{\beta+\left(\frac{n\pi}{2}\right)^2}{\frac{1}{\sqrt{2}}=\left\{\left(\frac{\pi a}{2\lambda}\right)^2+\left(\frac{\pi a}{2}\right)^2 f^2\right\}^{1/2}}$ unit : (g/2ωa) • A
1	1OO	1	0.7502743	1.1987862	76.317103	1.178528	76.878662
		2	0.4092318	1.3537094	43.089909	1.285640	44.962134
		3	0.2655595	1.3890564	29.476694	1.251419	32.453121
		4	0.1882130	1.4016333	22.307686	1.182577	26.251658
	10	1	0.6801818	1.2399003	7.893450	1.068427	8.480099
		2	0.3043081	1.3810854	4.396131	0.956012	6.046482
		3	0.1635824	1.4047209	2.980910	0.770864	5.268435
		4	0.0999305	1.4106786	2.245164	0.627878	4.944372
	1	1	0.4078598	1.3541234	0.862062	0.640665	1.414214
		2	0.1301072	1.4082159	0.448249	0.408744	1.414214
		3	0.0609401	1.4129000	0.299827	0.287174	1.414214
		4	0.0349375	1.4137819	0.225010	0.219519	1.414214
	1/10	1	0.1498877	1.4062481	0. 089525	0. 235443	0.384822
		2	0.0429605	1.4135609	0.044995	0.134969	0.428299
		3	0.0196559	1.4140770	0.030008	O.092626	0.438456
		4	0.0111731	1.4141694	0.022507	0.070203	0.442214
	1/100	1	0.0482519	1.4133902	0.008998	0.075794	0.119539
		2	0.0136470	1.4141477	10.004501	0.042873	0.134827
		3	0.0062283	1.4141998	0.003001	0.028461	0.138372

contd...

$\left(\frac{\pi a}{2\lambda}\right)$	β	n	$\dfrac{\left(\frac{\pi a}{2\lambda}\right)}{\frac{1}{\sqrt{2}}=\left\{\left(\frac{\pi a}{2\lambda}\right)^2+\left(\frac{\pi a}{2}\right)^2 f^2\right\}^{1/2}}$ (Unit : A)	$\dfrac{\left(\frac{\pi a}{2\lambda}\right)\cdot f}{\frac{1}{\sqrt{2}}=\left\{\left(\frac{\pi a}{2\lambda}\right)^2+\left(\frac{\pi a}{2}\right)^2 f^2\right\}^{1/2}}$ (Unit : A)	$\dfrac{\beta f}{\frac{1}{\sqrt{2}}=\left\{\left(\frac{\pi a}{2\lambda}\right)^2+\left(\frac{\pi a}{2}\right)^2 f^2\right\}^{1/2}}$ unit : (g/2ωa) • A	$\dfrac{\left(\frac{\pi a}{2\lambda}\right)\cdot\frac{n\pi}{2}}{\frac{1}{\sqrt{2}}=\left\{\left(\frac{\pi a}{2\lambda}\right)^2+\left(\frac{\pi a}{2}\right)^2 f^2\right\}^{1/2}}$ unit : (g/2ωa) • A	$\dfrac{\beta+\left(\frac{n\pi}{2}\right)^2}{\frac{1}{\sqrt{2}}=\left\{\left(\frac{\pi a}{2\lambda}\right)^2+\left(\frac{\pi a}{2}\right)^2 f^2\right\}^{1/2}}$ unit : (g/2ωa) • A
		4	0. 0035373	1.4142091	0.002250	0.022225	0.139681
1/10	100	1	0.0887614	1.4114252	89.854127	0.139426	90.951467
		2	0.0429246	1.4135619	44.995074	0.134852	47.161108
		3	0.0271412	1.4139531	30.005016	0.127900	33.168318
		4	0.0190558	1.4140852	22.505865	0.119731	26.578693
	10	1	0.0804691	1.4119224	8.988577	0.126400	10.032399
		2	0.0319191	1.4138533	4.500435	0.100277	6.342198
		3	0.0167187	1.4141147	3.000845	0.078785	5.384540
		4	0.0101175	1.4141773	2.250733	0.063570	5.005967
	1	1	0.0482519	1.4133902	0.899792	0.075794	1.673088
		2	0.0136470	1.4141477	0.450137	0.042873	1.483379
		3	0.0062283	1.4141998	0.300103	0.029350	1.445380
		4	0.0035373	1.4142092	0.225078	0.022225	1.431831
	1/10	1	0.0177325	1.4141024	0.090024	0.027854	0.455264
		2	0.0045062	1.4142064	0.045016	0.014157	0.449246
		3	0.0020089	1.4142121	0.030011	0.009467	0.448119
		4	0.0011312	1. 4142131	0.022508	0.007108	0.447723
	1/100	1	0.0057085	1. 4142015	0.009003	0.008967	0.141421
		2	0.0014314	1.4142128	0.004502	0.004497	0.141421
		3	0.0006366	1.4142134	0.003001	0.003000	0.141421

contd...

$\left(\frac{\pi a}{2\lambda}\right)$	β	n	$\dfrac{\left(\frac{\pi a}{2\lambda}\right)}{\frac{1}{\sqrt{2}}=\left\{\left(\frac{\pi a}{2\lambda}\right)^2+\left(\frac{\pi a}{2}\right)^2 f^2\right\}^{1/2}}$ (Unit : A)	$\dfrac{\left(\frac{\pi a}{2\lambda}\right)\cdot f}{\frac{1}{\sqrt{2}}=\left\{\left(\frac{\pi a}{2\lambda}\right)^2+\left(\frac{\pi a}{2}\right)^2 f^2\right\}^{1/2}}$ (Unit : A)	$\dfrac{\beta f}{\frac{1}{\sqrt{2}}=\left\{\left(\frac{\pi a}{2\lambda}\right)^2+\left(\frac{\pi a}{2}\right)^2 f^2\right\}^{1/2}}$ unit : (g/2ωa) • A	$\dfrac{\left(\frac{\pi a}{2\lambda}\right)\cdot\frac{n\pi}{2}}{\frac{1}{\sqrt{2}}=\left\{\left(\frac{\pi a}{2\lambda}\right)^2+\left(\frac{\pi a}{2}\right)^2 f^2\right\}^{1/2}}$ unit : (g/2ωa) • A	$\dfrac{\beta+\left(\frac{n\pi}{2}\right)^2}{\frac{1}{\sqrt{2}}=\left\{\left(\frac{\pi a}{2\lambda}\right)^2+\left(\frac{\pi a}{2}\right)^2 f^2\right\}^{1/2}}$ unit : (g/2ωa) • A
		4	0.0003581	1.4142135	0.002251	0.002250	0.141421
1/100	100	1	0.0088939	1.4141855	90.029852	0.013971	91.133740
		2	0.0042946	1.4142071	45. 015608	0.013492	47.184755
		3	0.0027147	1.4142109	30.010487	0.012793	33.175711
		4	0.0019058	1.4142122	22.507886	0.011975	26.582026
	10	1	0.0080630	1.4141902	9.003015	0.012665	10.052503
		2	0.0031935	1.4142100	4.501570	0.010033	6.345379
		3	0.0016722	1.4142126	3.001052	0.007880	5.385740
		4	0.0010119	1.4142132	2.250790	0.006358	5. 006595
	1	1	0.0048349	1.4142052	0.900311	0.007595	1.676441
		2	0.0013654	1.4142129	0.450158	0.004289	1. 484123
		3	0.0006230	1.4142135	0.300105	0.002936	1.445702
		4	0.0003538	1.4142135	0.225079	0.002223	1.432011
	1/10	1	0.0017768	1.4142124	0.090032	0.002791	0.456177
		2	0.0004508	1.4142135	0.045016	0.001416	0.449471
		3	0.0002009	1.4142136	0.030011	0.000947	0.448218
		4	0.0001131	1.4142136	0.022508	O. 000111	0.447779
	1/100	1	0.0005720	1.4142135	0.0090032	0.0008985	0.141705
		2	0.0001432	1.4142136	0.0045016	0.0004499	0.141492
		3	0.0000637	1. 4142136	0.0030011	0.0003000	0.141453
		4	0.0000358	1.4142136	0.0022508	0.0002251	0.141439

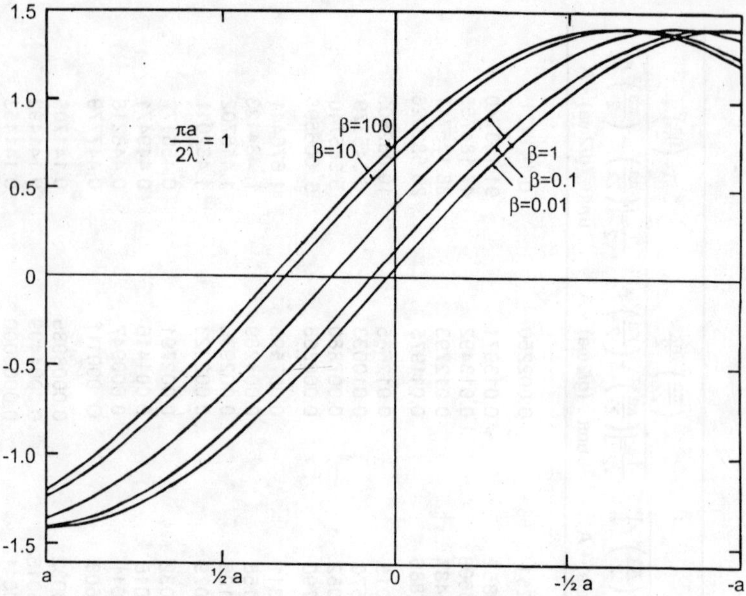

Fig. 6.12 : Cross-canal variation of wave amlitude.

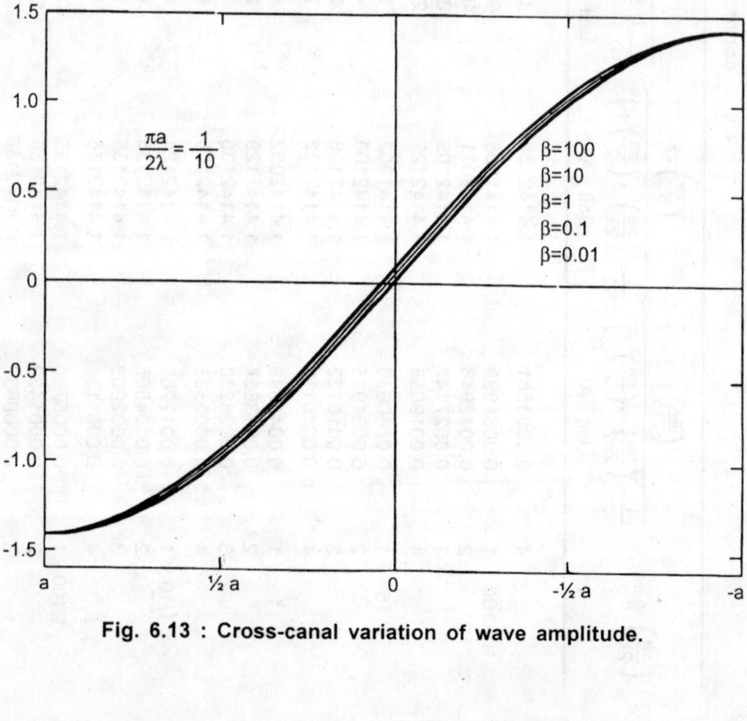

Fig. 6.13 : Cross-canal variation of wave amplitude.

canal, however broad the canal may be. Under this assumption we can show that the velocity of propagation of Kelvin waves is exactly (gh), where h is depth of the canal supposed to be uniform, and g the acceleration of gravity. On the contrary, we can by no means have the velocity of propagation equal to $\sqrt{(gh)}$ in Pointcaré waves.

Although it is almost 60 years since Poincaré published his result in "Mécanique Céleste "it does not appear to the author if any satisfactory manifestation of the behavior of Poincaré waves is ever given, except that Albert Defant gave some accounts on this type of waves.

The Numerical computations on Poincaré waves can be studied as follows.

Chapter Seven
The Positioning of Submersibles and Surface Ships

7.1 Introduction

The propulsors installed on today's small submersibles look like a throwback to the early part of the nineteenth century. The reasoning behind many of these designs bears a marked similarity to that applied more than 100 years ago.

Much of this reasoning is based on the fact that it takes very little force to move a small vehicle through the water at slow speeds. Vehicle shapes are such that high speeds wild require a prohibitive amount of power, and at low speed the variation in propulsive efficiency from poor to good does not seem to have any pronounced effect. The designer of these vehicles is more concerned with problems of environment and safety than with those of gaining a little more speed or range.

As more and more use is made of small submersibles for bottom exploration and for recovery of sunken objects, however, it is gradually being recognized that although speed may be unimportant, the control of the vehicle is extremely important. Paralleling the need for good control is the need to be able to stay on the job for as long as possible. The working period is often limited by power storage capacity, which in turn is a direct function of the efficiency of control-force generation. Thus the means to develop full control force with a minimum of power expenditure is being more frequently considered in small-submersible design.

The subject being considered is *positioning control* as opposed to the more familiar *maneuvering control* usually employed in waterborne vehicles. In order for one to have positioning control of a submersible in any single degree of freedom, it must be possible to develop either a force or a couple of controllable direction and magnitude in a single plane. This must be done without developing extraneous forces or couples in other planes.

A single-screw surface ship with a rudder behind its propeller can be discussed to show the difference between positioning control and maneuvering control. This ship has maneuvering control in the ahead-astern direction, or surge, and in rotating about a vertical axis, or yaw. It has positioning control, however, only in surge because, although it can develop fore-and-aft thrust in a unique manner, it cannot develop the turning moment for yaw control without also producing thrust.

Altogether there are six degrees of freedom in which it may be desirable to control a submersible. These are the translational motions of surge, sidle, and heave, and the rotational motions of roll, pitch, and yaw. Various conventions may be adopted for the axes and co-ordinates to which these motions are related, but for the purposes of this discussion, it is convenient to use the rigid-body or ship-axes system.

In this system the three orthogonal axes pass through the center of gravity of the vehicle. The x-axis extends from the bow to the stern, lying in the principal planes of transverse and vertical symmetry. The y-axis is perpendicular to the x-axis and lies in the principal plane of transverse symmetry.

Roll moment –	produces rotation about the x-axis
Pitch moment –	produces rotation about the y-axis
Yaw moment –	produces rotation about the z-axis

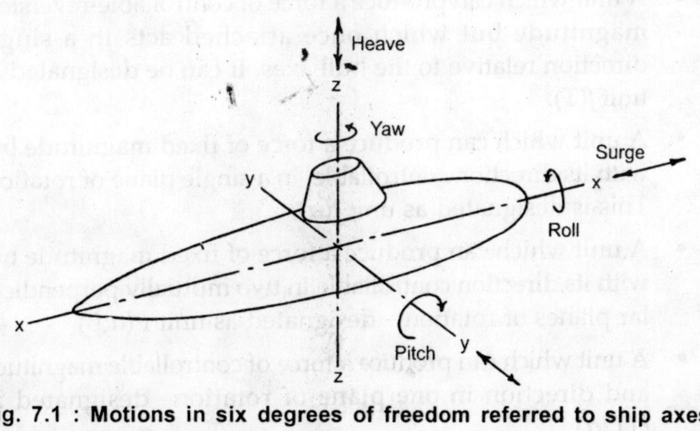

Fig. 7.1 : Motions in six degrees of freedom referred to ship axes.

The z-axis is mutually perpendicular to the x- and y-axes and lies in the principal plane of vertical symmetry.

Since the axes are fixed with respect to the vehicle, they are convenient for describing the forces and moments acting on the hull, but not for discussing motions which may result from the application of these forces and moments. It can be assumed, however, that when these forces and moments are applied to a neutrally buoyant body, a translation or rotation will take place in the direction of the force or moment and that the rate of motion will be proportional to the magnitude of the force or moment.

The forces and moments can then be categorized by the vehicle motions which they will produce as follows:

Surge force – produces translation along the x-axis

Sidle force - produces translation along the y-axis.

Heave force - produces translation along the z-axis.

7.2 The Characteristics of Submersible Hulls

Before we get involved in the capabilities of specific types of propulsion devices or in the characteristics of submersible hulls, the subject of positioning control can .be viewed in general terms. Consider the hull to be a sphere without mass or resistance on or in which can be spotted units of force. Certain characteristics can be assigned to these units of force which can later be related to actual force-producing elements and their location on a realistic vehicle. Five types of force units are involved:

- A unit which can produce a force of controllable-reversible magnitude but which once attached acts in a single direction relative to the hull axes. It can be designated as unit $f(T)$.
- A unit which can produce a force of fixed magnitude but with its direction controllable in a single plane of rotation. This is designated as unit $f(\alpha)$.
- A unit which can produce a force of fixed magnitude but with its direction controllable in two mutually perpendicular planes of rotation – designated as unit $f(\alpha,\beta)$
- A unit which can produce a force of controllable magnitude and direction in one plane of rotation– designated as $f(T, \alpha)$.

- A unit which can produce a force of controllable magnitude and direction in two planes of rotation – designated as unit f (T, α,β).

Positioning Control Ability of a Hypothetical Submersible

Number of Units	Type of Units	Degrees of Freedom in Translation and Rotation		Degrees of Freedom in Translation and Rotation	
1	f(T)	1	0	–	–
2	f(T)	2	0	1	1
3	f(T)	3	0	2	1
4	f(T)	3	1	2	2
5	f(T)	3	2	–	–
6	f(T)	3	3	–	–
1	f(α)	No positioning Control			
2	f(α)	1	1	–	–
3	f(α)	2	1	1	2
4	f(α)	2	2	1	3
5	f(α)	2	3	–	–
6	f(α)	3	3	–	–
1	f(α,β)	No positioning Control			
2	f(α,β)	2	2	–	–
3	f(α,β)	3	3	–	–
1	f(T,α)	2	–	–	–
2	f(T,α)	3	1	2	2
3	f(T,α)	3	3	–	–
1	f(T,α,β)	3	–	–	–
2	f(T,α,β)	3	2	–	–
3	f(T,α,β)	3	3	–	–

The above table shows the number and type of force units needed to obtain varying amounts of positioning control of this hypothetical submersible. This positioning control ability is expressed in terms of the number of degrees of freedom in which a given number of units of each type can produce translational forces or rotational couples or both.

A few examples may illustrate how some of these tabulated values were derived.

First, consider the case of three f (T) force units which deliver a controllable-reversible force only. If these three units are installed at the center of the sphere with their thrusts directed in mutually perpendicular directions, they produce forces for surge, sidle, and heave. Since each unit force can be reduced to zero, they can exercise unique control in any one of these degrees of freedom. By varying the proportions of unit force magnitude, the translation control can be applied in any direction.

Since the hull has been assumed to have no mass or resistance, a pure moment, or force couple, can be applied to it only by a pair of force units which are producing forces of equal magnitudes in opposite directions. These can be pictured as installed diametrically to each other on the spherical hull. Single force units can be pictured as installed at the center of the hull unless they are a part of a force-couple pair.

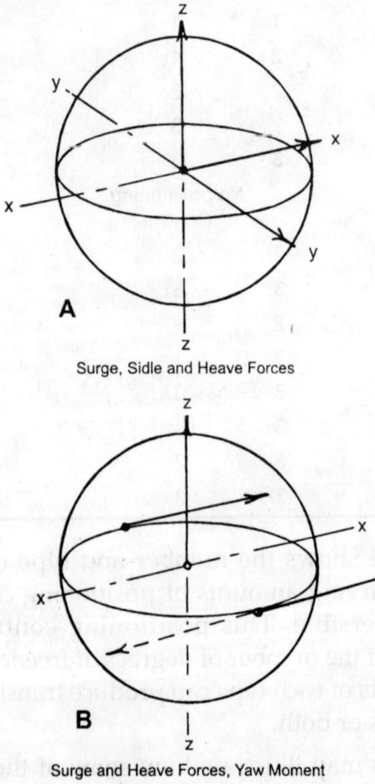

Fig. 7.2 : Combinations of three variable force nonrotatable units.

Now let us try to understand about three force units. These units are shown installed in two different ways in Figure. In Figure the three forces are directed toward the center in their neutral position, and all three rotate in the same plane. By swinging any pair in opposite directions, a translational force imbalance will develop to move the hull in the direction of the third unit. By proper control of all three, any combination of surge and sidle forces can be produced. If all three are rotated in the same direction, a yaw moment will develop.

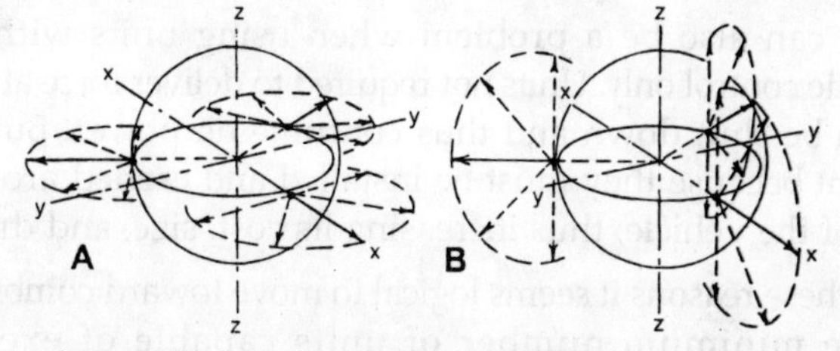

Fig. 7.3 : Combinations of three units with constant force rotating in one plane.

The three units are installed in the same location but with vertical planes of rotation. Swinging all three up or down together will provide a heave force. By controlled swinging of various units in different directions, any combination of roll and pitch moments can be applied to the hull.

The following diagrams indicate the force vectors rotated to give surge force sidle force roll moment and pitch moment.

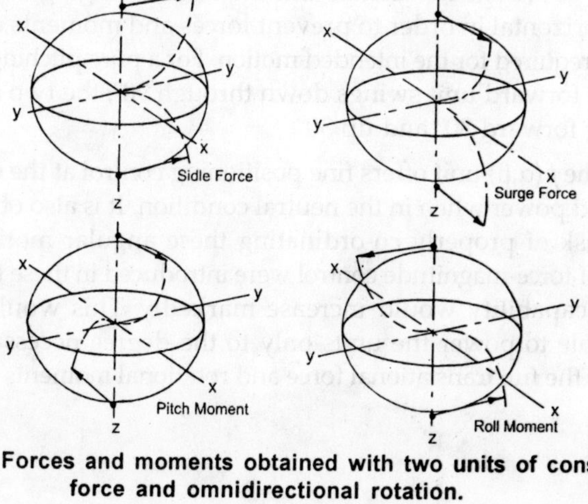

Fig. 7.4 : Forces and moments obtained with two units of constant force and omnidirectional rotation.

7.3 The Utilization of Force Units

The factor which must be considered, in the preliminary analysis, is the efficiency of utilization of the force units. Those units in which only direction can be controlled must operate continuously in order to provide a force and moment balance. This means that they will be consuming power even when delivering no usable force. When a force or moment is being applied, its total magnitude may also be considerably less than the sum of the unit capabilities.

This can also be a problem when using units with force-magnitude control only. Units not required to deliver force at a given time can be shut down and thus consume no power. But this is inefficient because they must be installed and carried around for the life of the vehicle, thus increasing its cost, size, and drag.

For these reasons it seems logical to move toward combinations with the minimum number of units capable of exercising positioning control in the degrees of freedom when control is required. In addition, these combinations should be such that the total force capability of each unit is used to the maximum extent possible in each kind of motion.

When all three units are rotated in the x-y plane through 90° in the same direction, a yaw moment results which uses the full capabilities of the three units. However, with the arrangement shown, the maximum moments for pitch and roll can only take advantage of half the total force available. For a pure rolling moment, the after units are rotated to thrust aft with one swinging to 60° above the horizontal and the other swinging to 60° below the horizontal in order to prevent forces and moments other than those required for the intended motion. For a pure pitching moment, as the forward unit swings down through 90°, the two after units swing forward 30° and up 30°.

The $f(\alpha,\beta)$ unit offers fine positioning control at the expense of wasted power when in the neutral condition. It is also obvious that the task of properly co-ordinating these angular motions is not easy. If force-magnitude control were introduced in these three units, their capability would increase markedly. This would make it possible to power the units only to the degree necessary and to apply the full translational force and rotational moments of all three units.

The capability of three $f(T,\alpha)$ units lies some-where between that of the other two. Figure 9-6 shows one means of orienting these units to give positioning control in six degrees of freedom. The total translational force which can be developed is equal to two-thirds of the total available in the three units since one unit must be shut down in each case. To obtain a pure roll, pitch, or yaw moment, all three units can be run up to full force output, but in each case one unit is needed to counteract the unwanted forces produced by the other two.

For example, in order to create a yawing moment, the forward unit swings 45° down from the horizontal and the starboard unit swings 45° up from the horizontal. This gives a yawing moment to port, and the heave forces cancel out. There remains, however, a net surge-sidle force and a roll-pitch moment which must be balanced out by running the upper unit as indicated. Thus in developing rotating moments, slightly less than half the total available force is utilized. But this type of unit has the advantage that the force output can be varied and power vaste reduced to a minimum.

So from a simplified analysis of positioning control of small submersibles, the logical choice of force units boils down to using a minimum of three units which deliver force of controllable magnitude and direction in either one or two planes of rotation, if control is to be exercised in all six degrees of freedom with the least number of units. Control of force application in any direction will utilize the installed power more effectively than if its direction is controlled in only one plane of rotation.

Up to this point there has been no consideration of what these units of force will comprise or how they might be installed on a realistic hull which has mass, drag, and metacentric stability. These force units must be powered in some manner, and the magnitude and direction of the forces which they produce must be controlled.

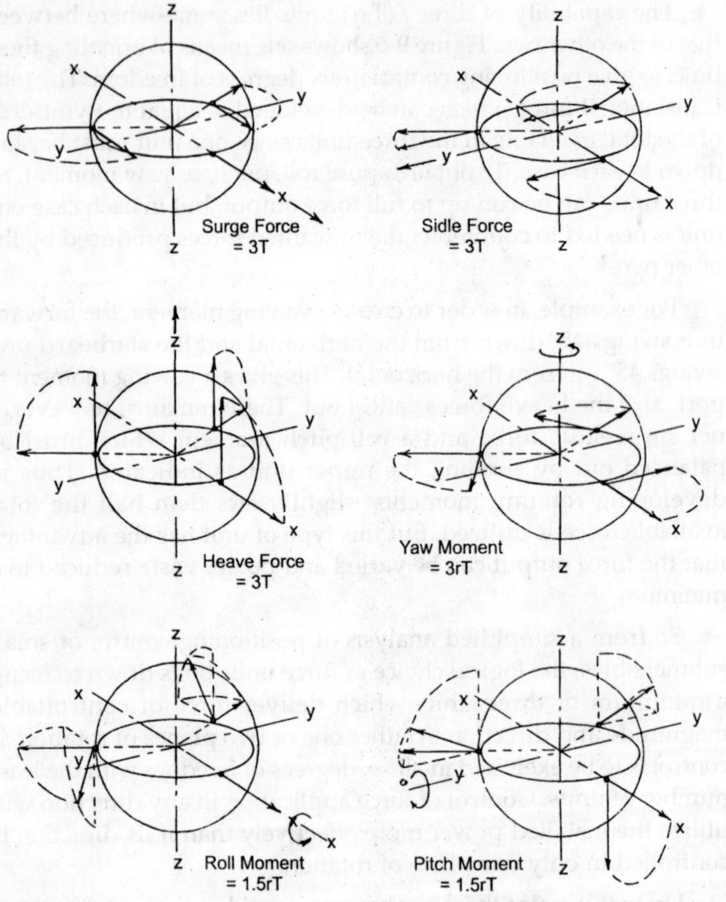

Fig. 7.5 : Forces and moments for motion in six degrees of freedom using three constant-force omnidirectional units.

The term *force unit* rather than *thrust unit* has been used purposely since force is available to a submersible from a source other than a propulsor. In a vertical direction, with respect to fixed earth axes, the buoyancy of the submersible can be varied to provide a force of controllable magnitude. Within reasonable limits, this force can be divided between multiple units to provide rotating moments as well. A single variable ballast tank provides a heave force, and two additional tanks can provide roll and pitch moments as well.

Normally, submersibles are fitted with trim and ballast systems for reasons other than positioning control. As discussed here, the buoyancy force units can be considered separately from these installed systems in order that they may be, compared rationally with propulsive force units.

A realistic value for the dead-pull thrust of a typical propulsor is on the order of 25 pounds per horsepower. To duplicate this specific thrust, a buoyancy force unit would require a volume of 0.4 cubic feet of sea water, which is a cube of about nine inches on a side. This is smaller than a one-horsepower motor-driven propeller which could deliver an equivalent thrust. Therefore, a buoyancy force unit is well worth considering as a replacement for a propulsive force unit of the type which has controllable force magnitude but acts in a fixed direction.

The usual source of power for a small submersible is a bank of storage batteries supplying direct current at low voltage. An alternate power source used for some functions is a high-pressure compressed-air supply. Both of these sources have the capability of delivering power at a rate controllable from very slow to very rapid. Both also have a total capacity in hours which is proportional to the rate of power drain. Direct current from the batteries may be converted to alternating current to drive submersible motors external to the pressure hull. Submersible air motors may be driven directly from supply tanks.

Although it is possible to house propulsor drive moters within the pressure hull, a sealing problem is involved where the drive shaft penetrates the hull. If the propulsor is to have thrust-direction control as well as thrust-magnitude control, an additional problem is encountered with submerged gearing and control elements. Generally it has been satisfactory to use submersible motors to drive propulsors directly. When directional control of thrust is required, the entire motor-propulsor assembly is rotated.

There are some practical limitations in rotating a motor-propulsor assembly. When an electric motor is used, the electric connections from the fixed hull to the rotating motor must pass through one or more submerged slip ring assemblies if 360° rotation without unwinding is needed. An alternative approach is to limit the rotation and use flexible cable, but this makes the control more complex. Similarly when air motors are used, either a rotating seal must be installed or flexible hose employed and rotation restricted.

In some small submersibles it has been expedient to install a number of unidirectional propulsors with the only control of thrust magnitude being an on-off switch. Obviously this arrangement does not permit fine positioning control of the vehicle and requires an excessive number of propelling units. A system of this sort is highly inefficient and is not recommended for submersibles which are intended to perform difficult underwater tasks.

A final practical consideration in propulsor selection is susceptibility to damage. When a submersible is working around sunken hulks, near reefs or outcroppings, or in areas where there are cables or hoses, the propulsor must be well protected. Damage to the propulsor itself is always a possibility. In addition, it might become entangled so as to prevent the submersible from surfacing. Generally, propulsive efficiency is inversely proportional to the degree of protection of the propulsor. Thus, for any design, a compromise must be made between what is desirable for positioning control and what is practical from an operational standpoint.

7.4 The Types of Propelling Devices

There are many kinds of propelling devices which could conceivably be used for positioning control of submersibles.

The most efficient means of obtaining propulsive thrust is to use a conventional screw propeller. When used at its designed rpm and speed of advance, it has yet to be beaten by any other propulsive device. It does have some limitations, however, as a submersible positioning unit. Its efficiency is less when operating at other than design conditions. Thrust may be varied either by changing rpm or blade pitch. A controllable-pitch propeller has an extended range of maximum efficiency, but the mechanism is complex.

An open-screw propeller on a submersible lacks the protection desired for many operations. Restricting its diameter helps to some extent, but its efficiency decreases. This situation improves if the propeller is fitted with a nozzle to regain the lost efficiency and also to afford protection. The efficiency gain is, however, in only one direction of operation with a corresponding loss of efficiency in reverse. Another approach is to install the propeller in a tunnel through the hull. This gives the needed protection but at reduced efficiency. In addition, the opportunity to control thrust direction is lost.

Control of screw propeller thrust direction can be attained by rotating its axis in one plane or in mutually perpendicular planes. This technique is subject to the practical limitations discussed earlier, but rotation of the axis in one plane has been used successfully in many applications. The thrust direction of propellers in nozzles has also been controlled by rotating the entire

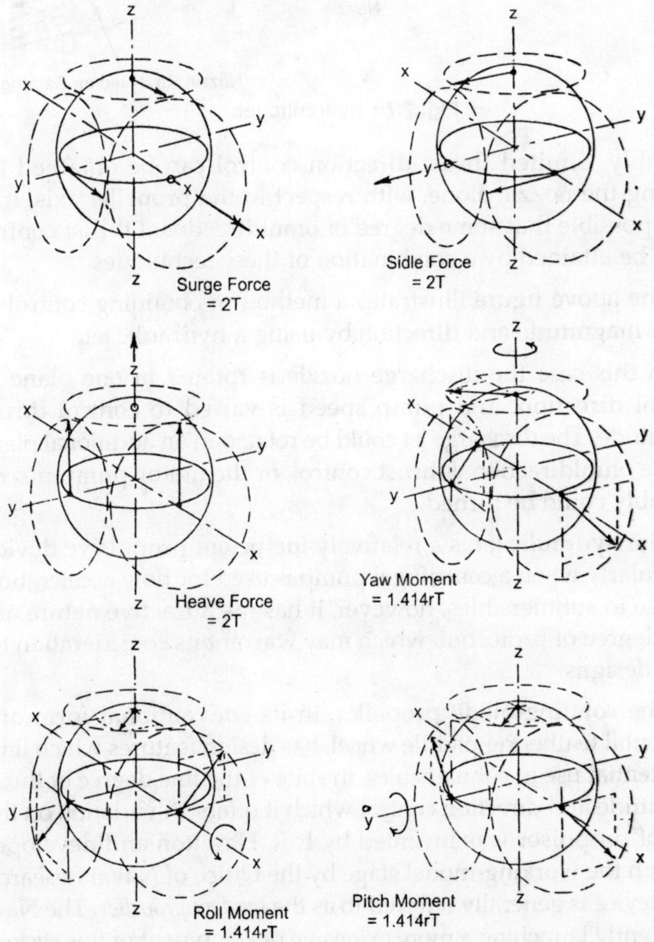

Fig. 7.6 : Forces and moments for motion in six degrees of freedom using three variable force units rotatable in one plane.

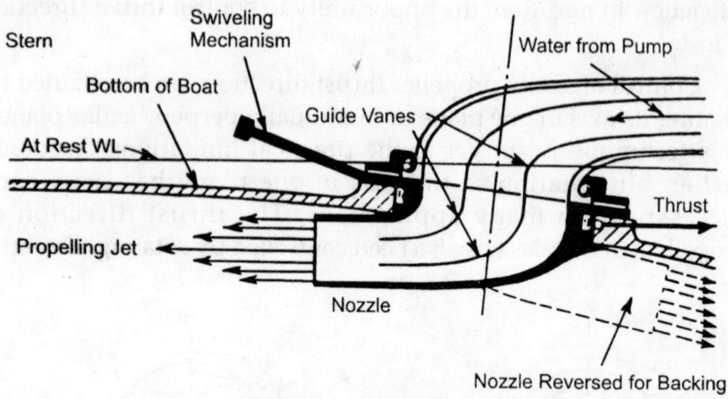

Fig. 7.7 : Hydraulic jet.

assembly. Limited thrust-direction control can be obtained by rotating the nozzle alone, with respect to the propeller axis. It is quite possible that some degree of omnidirectional thrust control could be attained by a combination of these techniques.

The above figure illustrates a method of obtaining control of thrust magnitude and direction by using a hydraulic jet.

In this case the discharge nozzle is rotated in one plane to control direction, and pump speed is varied to control thrust magnitude. The discharge jet could be rotated in an additional plane to give omnidirectional thrust control, or the motor-pump-nozzle assembly could be turned.

The hydraulic jet is a relatively inefficient propulsive device, particularly when a centrifugal pump is used for flow acceleration. For use in submersibles, however, it has the attractive nature of a high degree of protection which may warrant its consideration for some designs.

The rotating-blade propeller, in its conventional form of a horizontal feathering paddle wheel, has design features which limit its potential use in submersibles, in spite of the fine degree of thrust magnitude and direction control which it offers. A variation on this type of propulsor was invented by F. R. Haselton and developed through the working-model stage by the Office of Naval Research. This device is generally referred to as the *tandem propeller*. The Navy is currently launching a more extensive development of this system.

The fore-and-aft blade rings are individually driven and are external to the pressure hull. The pitch of the blades remains fixed for forward propulsion but can be cyclically varied to deliver thrust in any direction. The two units at the extremities of the vehicle can provide translational thrust in any direction and thrust couples to rotate the vehicle in any angular motion.

This pair of propulsors falls into the category of the two $f(T, \alpha, \beta)$ units discussed earlier. With properly co-ordinated blade motions, it very nearly approaches the ultimate in positioning control for submersibles. As shown in Figure 9–8, this tandem propeller installation can give unique control in all of the six degrees of freedom except roll. Although roll can be controlled, it cannot be done without inducing other motions as well.

There has been some concern about the vulnerability of the blade rings. A surrounding nozzle affords some protection, bet even this may not be sufficient for all applications. In addition, the drive system is admittedly complex, but the advantages in positioning controllability may well outweight these practical considerations.

7.5 The Positioning-Control Criteria

The force-producing elements necessary to dynamically position a submersible and to control its motions in all six degrees of freedom are available.

But it is also apparent that the number and complexity of the force units increases in proportion to the degree of control to be exercised. Therefore, it behooves the designer and the prospective operator to carefully list the vehicular motions necessary to fulfill the mission.

The next step in establishing positioning-control criteria is to determine the absolute and relative responsiveness desired in each degree of freedom to be controlled. These determinations, of course, involve some consideration of the vehicle configuration and some knowledge of propulsor capabilities as well as the submerged endurance requirements for the vehicle. They also require some knowledge of the operating environment in terms of horizontal and vertical currents which may act on the vehicle and the attitude which the vehicle may have to assume relative to these currents.

Once these factors are known or estimated, one can calculate the resistance and inertial forces for motion in each degree of freedom as a function of applied forces and moments. At this point one should also consider the possibility of coupled motions resulting from configuration anomalies.

Finally, these studies should result in a specification of the forces and moments required of the positioning-control force units, together with a limitation on the allowable power demand. These requirements must be firmly established before an intelligent selection can be made of the types of force units required and their location on the vehicle.

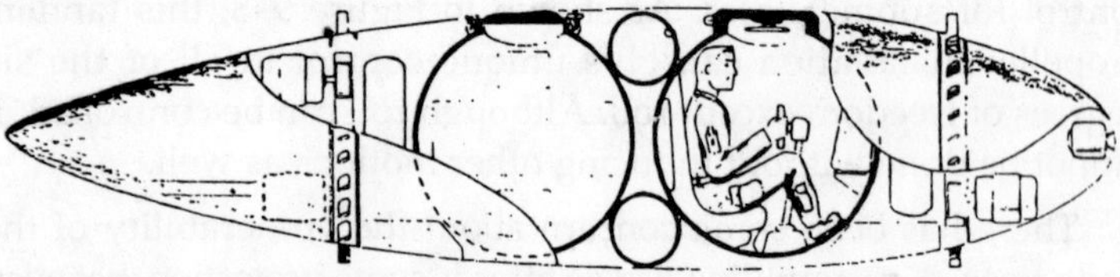

Fig. 7.8 : Haselton's proposed tandem propeller installation.

There are numerous types of propulsion systems which have been invented to provide controllable directional thrust. For this particular project, however, it was essential to consider only the few which had been developed for commercial use and had proven capabilities. This narrowed the field to the air screw, the hydraulic jet, the steering screw, and the vertical axis propeller. To augment the low-speed thrust and controllability of some of these types, the Kort nozzle and the controllable-pitch propeller were also included.

Air-screw propulsion for a waterborne vehicle refers to a method of obtaining propulsive thrust by using aircraft propellers driven by engines mounted topside. Air propulsion has been used in recent years to determine the resistance characteristics of full-scale ship hulls.

An interesting experiment has been conducted recently by the U.S. Navy in connection with the mine defense program. Large turbo-prop engines were installed on the gun mounts of a converted Liberty ship, and the underwater screw propeller was removed. The gun mounts could be rotated to direct the thrust of each propeller. By controlling propeller pitch and direction, a single man could maneuver the ship with ease.

There are some types of waterborne craft for which air-screw propulsion is the optimum solution. In swamp areas any propulsive device which projects below the hull, or in fact any device which requires water to act upon, is impractical. In these areas air screws mounted on tripods at the stern of a boat have worked successfully. Craft of this type have operated for many years in the Florida Everglades and in the Louisiana bayou country.

Despite these successful applications, air-screw propulsion is generally inefficient and requires expensive fuel.

For dynamic positioning of the *Cuss I*, the air screw offered some advantages. The primary requirement for high thrust in controlling the drill rig was under conditions of high wind velocity. Such conditions would contribute to higher efficiency of the air screw than normally anticipated. With rotatable self-contained engines and controllable-pitch air screws, a fine degree of controllability could be achieved.

However, the disadvantages of the air screw appeared to outweigh these few advantages. Unless the wind velocity were higher by a factor of three or four than that expected, the efficiency of the units would still be low. Coupled with the inefficient operation would be high fuel cost, noise and the danger to deck personnel of high-speed rotating blades.

Propulsion by a liquid jet emitting from the hull is one of the oldest forms of mechanically imparting motion to water craft. Interest in hydraulic jets has continued. Yet, in spite of the years of experimentation there has not yet been a successful application, except for specific purposes.

Although the hydraulic jet in many ways appeared attractive for the *Cuss I* installation, a unit has not been developed which takes advantage of all its favorable characteristics. The time and funds available for our project did not permit launching a development program. Proven methods of propulsion had to be utilized.

One of the simplest methods of producing thrust in any desired direction is the steering screw; that is, a screw propeller which can be rotated about a vertical axis which is perpendicular to the propeller rotation axis.

In the *Cuss I* installation, it was evident that 360° variation of the thrust direction was needed. This requirement called for a vertical shaft with a level gear drive to the propeller. Such an installation was used in the U.S. torpedo boat *Alarm* as early as

1879 and has been developed to a high degree of efficiency and reliability in the modern outboard motor.

Generally, the outboard motor does not approach the thrust requirements of *Cuss I.* There is however, a well-developed propulsive unit similar to the outboard motor which does provide the necessary thrust. This is the Murray & Tregurtha Harbormaster outboard propulsion and steering unit. The Harbormaster is a right-angle drive system with 360° rotation without recovery required. These units have been used extensively for barge and crane propulsion and are available in sizes from 20 to 1,200 horsepower. The usual prime mover for these units is a diesel engine, although the choice is optional. On some units the vertical shaft pivots about an axis on deck so that the entire propeller assembly can be swung up for repair and maintenance.

The Harbormaster has many advantages. It can be deck mounted and easily removed and replaced when maintenance is necessary. It is rugged and reliable – up to 55,000 hours of operation before major overhauls are required. Control systems can be worked out to give accurate directional control for multiple-unit installations.

The primary disadvantages of the Harbormaster unit are the lack of efficiency of a screw propeller in the 100% slip condition and the difficulty of closely controlling thrust over a wide range of magnitude and direction.

There are methods of overcoming these disadvantages. The Kort nozzle can be used to increase thrust greatly when the unit is operating at 100% slip or, in other words, operating at zero speed of advance. The controllable-reversible-pitch propeller could be used advantageously to provide a fine control of thrust magnitude and direction.

It would certainly be feasible to install a Kort nozzle on the Harbormaster propeller, but the problem of controlling thrust magnitude and direction is not so easily solved. Thrust direction is the fore-and-aft direction of thrust relative to the propeller rather than that which can be achieved by rotating the propeller disc. For fine control, it would be most desirable to alter the thrust gradually from a given magnitude in one direction through the zero thrust condition to a given thrust in the opposite direction. To do this with the Harbormaster unit requires lowering the speed to zero, rotating the unit 180°, and then increasing the speed again. For fine control when a small thrust is required, this would mean rapid rotation of the vertical shaft from one direction to the other.

The controllable-reversible-pitch, or CRP propeller, at first appears to be the obvious solution. Certainly such an installation would permit the necessary thrust control without complex controls for propeller speed and vertical shaft rotation. In addition, CRP propellers have been used successfully for many years. The difficulty which arises, however, is the incorporation of the control mechanism in the right-angle drive of the propeller. Although this could conceivably be done, it would require a long period of design and development which was not feasible for the *Cuss I* installation.

Another, more practical means of overcoming the thrust-control problem when multiple units are considered involves working pairs of units in opposition to each other. If two units are producing thrust in opposite directions, a slight increase in the thrust of one unit over the other can produce a very small thrust in the desired direction.

As for hydraulic jet propulsion, directional control of multiple propulsive devices would make such a system feasible. In fact if more than two units are installed, it is quite conceivable that a constant-speed drive could be used and thrust could be closely controlled by directional control of the propulsive units only.

One of the most unique and interesting propulsion devices which supplies a controllable, directional thrust is the vertical-axis propeller. The term *vertical-axis propeller* includes that wide variety of propellers which have been called horizontal paddle wheels, rotating-blade propellers, cycloidal propellers, and other descriptive terms. The term refers basically to those propellers having a vertical axis of rotation with a series of vertical blades disposed in a circle around the rotational axis which also have some form of rotation on their own axes.

In all modern types of vertical-axis propellers, the blades are spade-shaped hydrofoils. They are mounted on a large circular plate which fits neatly in a broad flat area on the underwater body of the ship. The circular rotor plate is barely submerged. A well is provided which extends above the waterline, thus preventing water from leaking into the ship. Small vanes are frequently provided around the periphery of the rotor so that water flowing into the well is pumped out, thus reducing frictional resistance around the rotor casing.

The German navy installed Voith-Schneider propellers in a group of 350-ton floating cranes which are faced with a positioning problem similar to that of the Mohole drilling ship. The only one of these now in service was brought to the United States and is presently working out of the Long Beach Naval Shipyard. This, the largest floating crane in the world, is fitted with three vertical-axis propellers, two aft and one forward. Each propeller is independently driven by a 700-horsepower electric motor. With a single operator at the controls, this huge crane can be maneuvered with ease, moving forward, backward, sideways, or diagonally. It can easily be held against a wind or tide while working alongside a pier or a ship. The personnel aboard are enthusiastic in their descriptions of its performance under all sorts of adverse conditions.

The vertical-axis propeller is gradually finding if its place in the marine propulsion field. It cannot compete with the screw propeller in ships designed for long ocean voyages, but when a high degree of maneuverability is required, the vertical-axis propeller is versatile and useful.

Because of its many assets and proven capability, the vertical-axis propeller appeared to be the ideal solution for dynamically positioning the *Cuss I.* Of the types available, only the Voith-Schneider units are produced in a series of sizes for ready selection. The units developed by Pacific Car & Foundry have been specially tailored for specific installations. Therefore an allowance for design time would be necessary. Figure 8-1 shows how three of these units could have been installed in *Cuss I.*

Although the vertical-axis propeller seemed to be the best technical solution to the positioning problem, practical considerations militated against its use for the experimental drilling program. A long lead time was required for delivery of three units from Germany, and the cost was high. Perhaps the major consideration, however, was the difficulty of installing these units in *Cuss I* and then removing them after the operation was completed. Extensive modification of the ship would be required, which did not appear feasible or economical for an operation of an estimated 30-day duration.

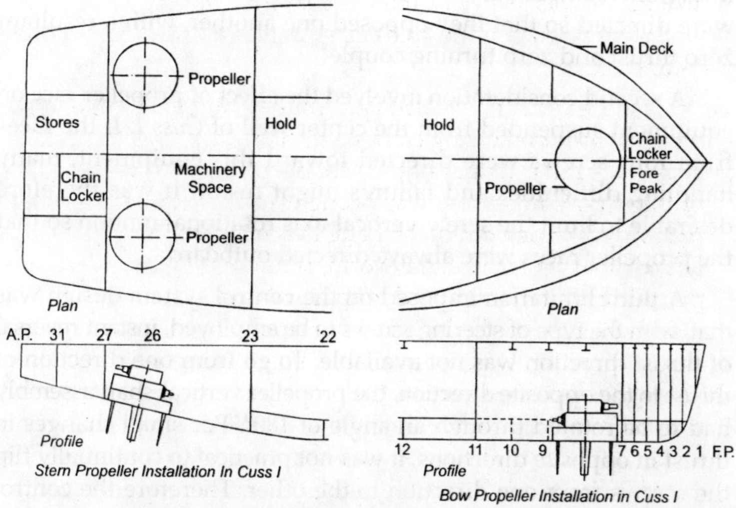

Fig. 7.9 : Three vertical-axis propeller units which could have been installed *Cuss I*.

7.6 The Complete Propulsion System

The function of the complete propulsion system is to hold the ship in a fixed position and head it into the prevailing swell regardless of the direction of wind and current.

This dictates that a combined thrust of the four units be resolvable into a single vector of any direction or magnitude, into a couple which would rotate the ship in either direction, or both concurrently. If possible, these resolved forces and moments should utilize the total thrust available from each of the propellers.

An optimum control system should be able to direct the propeller thrusts in these ways with variable magnitudes. There are, however, several other factors for which the control system design must provide.

First, the diesel engines which drive the steering screws cannot be started instantaneously to provide thrust of a given magnitude at the flick of a switch. For proper operation, the engines should be turning over, at least idling, continuously. Unless the control system is to be complicated by the unnecessary addition of a clutching arrangement, this means that the propellers will always be rotating and delivering mirlimal thrust. To keep the ship from moving while idling, it was necessary that the idling thrusts of the four screws were directed so that they opposed one another, with a resultant zero thrust and zero turning couple.

A second consideration involved the effect of propeller race on equipment suspended from the center well of *Cuss I*. If the races from four screws were directed toward this equipment, many handling difficulties and failures might result. It was therefore desirable to limit the screw vertical-axis rotational motion so that the propeller races were always directed outboard.

A third limitation imposed on the control system design was that, with the type of steering screws to be employed, instant reversal of thrust direction was not available. To go from one direction of thrust to the opposite direction, the propeller vertical-shaft assembly had to be rotated through an angle of 180°. For small changes in thrust in opposite directions, it was not practical to continually flip the screws from one direction to the other. Therefore the control system had to be designed to provide small thrust imbalances either by small alterations of the opposing thrust magnitudes or by small angular position changes or both.

For rotation of the ship, the screws must all turn away from the neutral position in the same direction of rotation. For the translational movements of the ship, opposite pairs of screws must operate in opposite directions of rotation.

Similarly, when the normal reactions of an operator were considered, two separate controls seemed desirable, one for rotation of the ship, and one for translation of the ship in any chosen direction. Thus from both the standpoint of vehicle movement and operator reactions, the control system should consist of two fundamental components, a rotational control and a translational control. These components must be interconnected to achieve combined ship movements.

This requirement led to a differential control system. Such a system has a master control unit, a differential control unit, and a follower or slave unit. If the master control unit is moved, either rotationally or linearly, and the differential control is held fixed, the slave unit will move in proportion to the movement of the master control.

Similarly, if the differential control is moved and the master control is held fixed, the slave unit will move an angular or linear amount proportional to the motion of the differential control. The position of the slave unit will always be proportional to the difference between the positions of the master and differential controls.

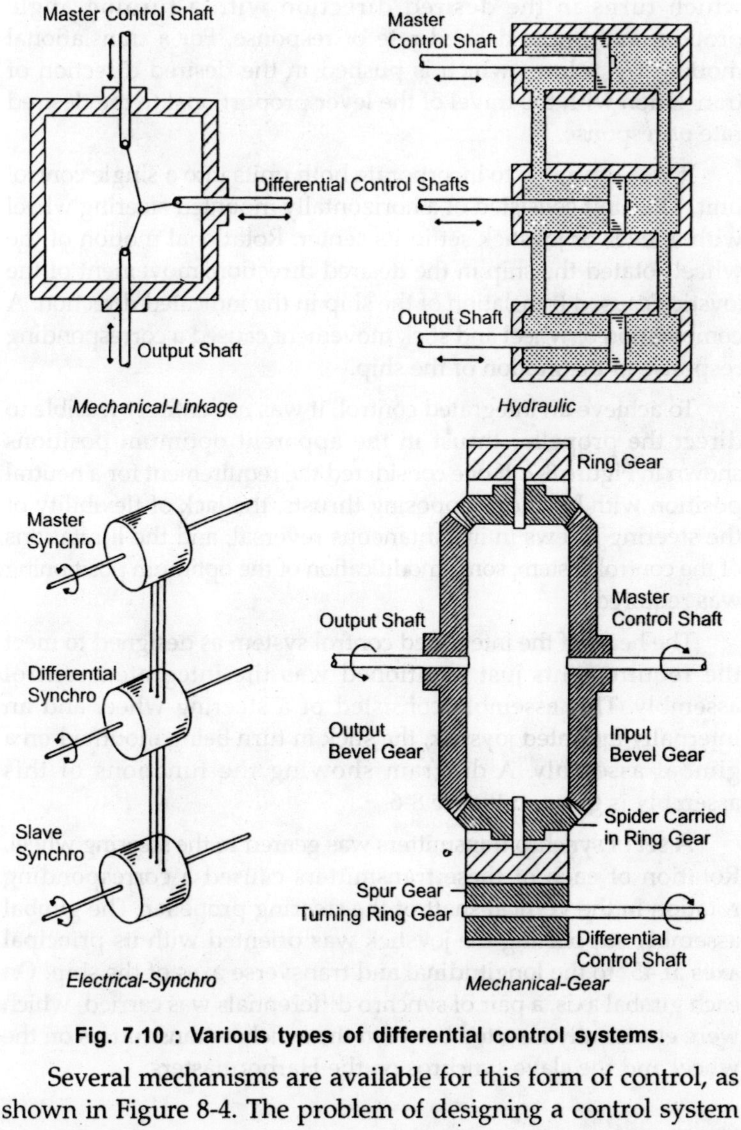

Fig. 7.10 : Various types of differential control systems.

Several mechanisms are available for this form of control, as shown in Figure 8-4. The problem of designing a control system for four steering screws thus became one of selecting one of these differential control methods and of evolving a combination of units which would give the desired response of the screws to the control operations.

The simplest control system for rotational motion is a wheel which turns in the desired direction with a turning angle proportional to the desired rate of response. For a translational motion it is a lever which is pushed in the desired direction of translation with the travel of the lever proportional to the desired rate of response.

It was desirable to incorporate both units into a single control unit. This unit consisted of a horizontally mounted steering wheel with a lever or joystick set in its center. Rotational motion of the wheel rotated the ship in the desired direction; movement of the joystick caused translation of the ship in the indicated direction. A combination of wheel and stick movement caused a corresponding response in the motion of the ship.

To achieve an integrated control, it was not entirely feasible to direct the propeller thrust in the apparent optimum positions shown in Figure 8-2. If one considered the requirement for a neutral position with balanced opposing thrusts, the lack of flexibility of the steering screws in instantaneous reversal, and the limitations of the control system, some modification of the optimum positioning was required.

The heart of the integrated control system as designed to meet the requirements just mentioned was the integrated control assembly. This assembly consisted of a steering wheel and an internally mounted joystick, the stick in turn being mounted on a gimbal assembly. A diagram showing the functions of this assembly is given in Figure 8-6.

A set of synchro transmitters was geared to the steering wheel. Rotation of each of these transmitters caused a corresponding rotation in the vertical shaft of the steering propeller. The gimbal assembly supporting the joystick was oriented with its principal axes at 45° to the longitudinal and transverse axes of the ship. On each gimbal axis, a pair of synchro differentials was carried, which were electrically inserted between the synchro transmitters on the wheel and the slave synchros on the Harbormasters.

If the joystick were on center and the steering wheel were rotated, the slave synchros on the Harbormasters would all rotate in the same direction away from the neutral position, thus imposing a rotating couple on the ship. With the wheel in the neutral position, movement of the stick in the plane perpendicular to the axes of differentials 1 and 4 would cause a right-hand rotation of differential 4 and a left-hand rotation of differential 1 and no rotation of 2 and 3. The corresponding motion of the forward-port slave and the after-starboard slave would be in opposite directions. The resultant thrust of the screws would be directed along the same line as the joystick actuation.

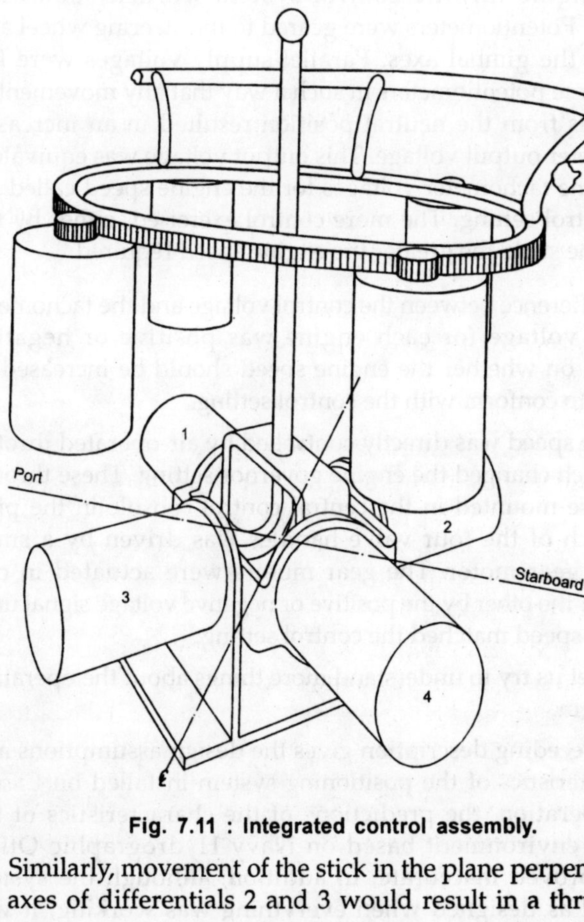

Fig. 7.11 : Integrated control assembly.

Similarly, movement of the stick in the plane perpendicular, to the axes of differentials 2 and 3 would result in a thrust in that direction. Movement of the stick in any intermediate plane would result in ship motion in the controlled direction.

In actual application, the synchros were tiny control transmitters, control differentials, and control transformers. The control transformers were geared to the vertical shafts of each of the four Harbormasters and produced error voltages corresponding to the difference between control position and propeller direction. The error signals were amplified, their phase sensed, and used to actuate rotation motor starters which rotated the vertical shafts of the Harbormasters until the error singnals were reduced to zero.

The engine throttle-control system worked somewhat differently. Potentiometers were geared to the steering wheel and to each of the gimbal axes. Parallel supply voltages were fed through these potentiometers in such a way that any movement of the controls from the neutral position resulted in an increased potentiometer output voltage. This output voltage was equivalent to the engine tachometer voltages for the engine speed called for by the control setting. The more control exercised, either by the wheel or the stick, the greater the engine speed required.

The difference between the control voltage and the tachometer generator voltage for each engine was positive or negative depending on whether the engine speed should be increased or decreased to conform with the control setting.

Engine speed was directly controlled by air-operated throttle valves which changed the engine governor setting. These throttle valves were mounted in the central control console in the pilot house. Each of the four valve handles was driven by a small reversible gear motor. The gear motors were actuated in one direction or the other by the positive or negative voltage signal until the engine speed matched the control setting.

Now let us try to understand more things about the operation of the system.

The preceding description gives the design assumptions and the characteristics of the positioning system installed on *Cuss I*. During operation, the predictions of the characteristics of the operating environment based on Navy Hydrographic Office statistics proved inaccurate. In addition, although the system performed as designed when everything was working, it was necessary to alter operating techniques because the system was not reliable.

The following account of the trip to the Guadalupe drilling site gives some idea of what a designer should anticipate from the ocean environment.

The *Cuss I* pulled away from the pier at National Steel and Shipbuilding Company, San Diego at 1300 hours on March 23, 1961, and made a few turns around the harbor area to check out all control systems. Since all controls operated satisfactorily, the *E.V. Turner* took the *Cuss I* in tow and headed for the Guadulupe test site.

On Friday, March 24, we maintained a heading of 270° to 290° because of the heavy swell which was running. A change of heading to the south during the previous night had caused the ship to roll excessively. Although the swell was still heavy during the evening, the course was changed to 140° in order to head toward Guadalupe Island. The roll was still substantial but seldom exceeded 20°.

On Saturday, March 25, the control systems were again checked; the engine connectors were opened and found to be corroded by salt water. The connectors were thoroughly cleaned with fresh water, contact cleaner, and carbo tetrachloride, and were packed with silicone grease. Except for erratic operation of the No. 3 engine tachometer, the control system then appeared to be satisfactory. About noon, the after engines were started to assist in increasing the speed of the tow. It was estimated that this increased the speed about one half knot.

On Sunday, March 26, the radar picked up Guadalupe Island at a bearing of 180 and a range of 42 miles.

The anemometer, which had been removed from the ship for repair during the La Jolla operation, was reinstalled. The wind-direction indicator appeared to work; but the wind-speed indicator needed further repair.

We arrived at the station area at 2130 hours, and the *E. V. Turner* cast loose the *Cuss I*. Under its own power, the *Cuss I* continued on a southerly course until clear of the location buoys. Amidst considerable confusion and heavy winds and seas, the *Cuss I* beat against the wind and swell. Position was held visually by means of the one lighted buoy and another buoy which could be seen intermittently with a searchlight. We attempted a steady position, but a strong current from the west kept pushing the ship off. Finally, we decided to hang on the one lighted buoy until morning.

On Tuesday, March 28, the current from the east seemed to drop off around 0400 hours. The engine speed was cut back to 1,300 rpm on the forward engines and 1,500 rpm on the after engines. A minor emergency occurred at 2200 hours when the fuel pressure on the after pair of engines dropped, indicating that the filters needed replacement. We discovered that it was possible to stop the engines one at a time and hold position while the filters were changed.

The wind continued at 25–30 mph with 10- to 12-foot waves most of the night. The ship rolled occasionally up to 11° due to cross swells which could not be avoided by changing the ship's heading. The No.1 engine high-temperature alarm went off, but a thorough check of the engine indicated it to be in good operating condition.

On March 29, Mate Garcia noted that the race of the propeller on the No.2 engine did not follow the position shown on the indicator synchro. The unit was then readjusted so that the indicator matched the apparent propeller race. Drilling continued throughout the day. The wind and sea began to abate around 1800 hours.

On Thursday, March 30, the inclinometer (Welex unit) was lowered to check it out as a positioning device. The indicator showed apparent movement but lagged considerably behind the ship movement as indicated by the radar and sonar and was very slow to return to its original position. The sonar had been giving good range information but poor azimuth information. The replacement of a rotary switch corrected this, and for the first time we had good sonar information.

By March 31, three radar buoys and three sonar buoys were operating, the maximum available during any part of our stay at Guadalupe.

A new drill string had been made up and lowered throughout the day and about 1630 hours, just as a touchdown was imminent, the steering wheel on the console froze.

Drilling operations were suspended while the trouble was investigated. A damaged bevel gear deep in the console was replaced to correct the situation. Then the No. 4 engine control system began performing erratically. We suspected that this was due again to a connector, but the engine could not be checked because high winds of 30–35 mph and eight-foot seas required almost total power. However, the decision was made to drill-in, and a depth of 200 feet into the bottom was reached by midnight.

Chapter Eight
The Air-Sea Interface

8.1 The Transition Techniques in Ocean

Today many interesting techniques have been developed to ease the transition from one floating vechicle to another or to a fixed structure.

In 1966, the Department of Defense invited industry to come up with designs and operating plans for a fleet of Fast Deployment Logistic Ships (FDLS). These ships would be deployed in range of potential trouble spots throughout the world and would be loaded with the vehicles and material to support a division of combat troops. When an emergency arose, they would be capable of speeding to the area and offloading their total cargo into landing craft some five miles off shore. One among many formidable design problems was that of carrying some of the landing craft, getting them into the water, and then transferring vehicular cargo from two forward holds into the landing craft.

The study reported herein was undertaken to develop a logical and efficient means of offloading landing craft and vehicles from holds 3 and 4 of the Fast Deployment Logistic Ship. Since the study was performed while the conceptual design was in progress, it was not possible to revise cargo-handling arrangements with each design change. The concepts which have been developed, however, include provisions for the most severe requirements for handling speeds, unit weights, and sea conditions. If less severe requirements are imposed, the cargo-handling system can readily be revised to meet them.

It was assumed that holds 3 and 4 are both 84 feet long with vehicles stowed on five intermediate decks and on the tank top. In each hold and in each deck, hatch openings are provided port and starboard which have a clear opening of 42 feet fore and aft by

17 feet athwartships, less the rounding of corners. Hatch covers are flush with the deck and will have vehicles stowed on them. Movement of vehicles within each tween deck space will be accomplished under their own power. The weight of vehicles to be offloaded will range up to 60 tons with maximum dimensions not exceeding the hatch cross section nor the maximum deck-to-coaming height of 11 feet, 8 inches.

A number of landing craft will be carried as deck cargo in holds 3 and 4. These will consist of a maximum of four LCM-8's and two LCL's. The cargo-handling gear, or auxiliary gear if necessary, is to be capable of offloading these landing craft and of then transferring the vehicular cargo from the FDLS to the landing craft alongside. It would be desirable to provide for offloading all landing craft either to port or to starboard to take advantage of the lee of the FDLS in adverse weather. Similarly, provision should be made for offloading vehicles from any hatch to either side of the ship.

Maximum speed in offloading boats and cargo is essential. The requirement is to unload all cargo and get it ashore in 20 hours in adverse weather conditions, up to sea state 5. Provision must also be made to rapidly reload the cargo and boats in the event of emergency evacuation of a beachhead.

In the design of cargo-handling rigging, structure, and machinery, the vertical moment should be kept to a minimum in order to maintain control of ship stability and roll characteristics. Weight of components is less important if the vertical location of the weight can be controlled to keep the center of gravity low.

The requirement for rapidly offloading boats and cargo in adverse sea conditions leaves no margin for loss of time. From the moment each vehicle is in position to be raised from the hold to the moment the landing craft moves away with the vehicle aboard, each step of the transfer must be made with precision. The operation must be conducted on a production-line basis. Each step in the transfer must be analyzed to make sure that it does not require a disproportionate amount of time and that it does not interfere with the accomplishment of any other step. Similarly, there is no margin for mishandling which would damage either a unit of cargo or one of the landing craft.

When the operation is viewed in this light, there are certain obvious principles which must be incorporated in the cargo-handling system. Complete control of each unit of cargo must be maintained from the instant it is lifted from the hold until it is firmly secured to the deck of the landing craft. Control of the landing craft must also be maintained during the loading process so that its relationship to the rest of the cargo-handling system is a determinant at all times.

Basically, each unit of cargo must be held within narrow limits of motion, relative to the ship structure and relative to the landing craft, from pickup to release. It must not move in any direction to an extent that would allow damaging contact. Throughout all phases of the transfer, pendulation of the load must be minimized in the design of the handling equipment. It would be improvident to rely on deck personnel with steadying lines as a substitute for mechanical control.

Use of personnel must be minimized throughout the operation. Contact with a unit of cargo should be restricted to attaching the unit to the hoisting device in the hold and to releasing the unit when it is deposited in the landing craft. Even in these operations, mechanical aids should be provided so that the skill of personnel is utilized rather than their strength.

Maintaining control of the relative position of the landing craft and the cargo-handling system implies that the landing craft is attached to the ship. Because of the large relative motion between the two in a seaway and the magnitude of the interacting forces involved, it is obviously impractical to secure the craft to the ship so that no relative motion exists. The range of relative motion must be restricted, however, so that the landing craft will not strike the side of the ship and so that the cargo can be guided into position in the well of the landing craft without damaging the cargo or the craft.

Compensation for the remaining relative motion between ship and landing craft must be provided in the cargo-handling gear. As a unit of cargo is moved from the hold up to the main deck and over the side, it must be controlled to move with the ship. As it is lowered into the landing craft, its motion must gradually make the transition from ship motion to boat motion so that when it reaches the deck of the landing craft, it is moving with the craft in all degrees of freedom. The total motion-adaptation system comprises the restriction which is imposed on the motion of the landing craft and the control exercised on the motion of the load in the transition between motion of the ship and motion of the landing craft.

When the various methods of cargo handling which were described previously are examined in light of the foregoing principles, the selection of a system which meets the FDLS specifications is considerably narrowed. Any arrangement which requires suspending units of cargo from the end of a single line or even a pair of lines cannot provide the essential control, either in raising the cargo from the hold or in lowering it into landing craft alongside. This eliminates considering single king posts and booms, burtoning rigs, rotating masts and booms, and any combinations of these methods.

To maintain complete control of a unit of cargo from the time it is picked up in the hold to its placement in the landing craft and to adapt the motion of the cargo to that of the landing craft in heavy seas, a minimum of four points of support is required. Furthermore the relative motion of these four support points in all degrees of freedom must be under control at all times.

Of the many types of cargo-handling systems available, only the overhead bridge crane or the gantry crane has these necessary features. None of these types of rigs have all of the features required to meet the exacting specifications for rapid offloading of vehicles into landing craft under extremely adverse sea conditions. It has, therefore, become necessary to select the applicable features of various cargo-handling systems, to devise methods of fulfilling requirements which existing systems do not cover, and to combine these in a conceptual design which provides for all the required functions.

8.2 The Structural Arrangement for Sliding Gantry

For the ship to meet Panama Canal transit requirements, all cargo-handling gear must be stowed within the maximum beam of the ship. Yet, to provide for controlled offloading of cargo into landing craft, the gantry frames must extend outboard from the main deck at side, about 38 feet. There are three alternative methods of achieving this extension.

The first method, which has been discarded for many practical reasons, would be to have a revolving gantry frame for each hold which would run fore and aft in the stowed position and would swing outboard on either side of the ship when revolved 90°. Each of these frames would have a total length of 180 feet.

A second method is to hinge the port and starboard gantry frames for each hold along lines inboard of the hatches and above the main deck. If a centerline bridge structure supports topping lifts, the outboard ends of the gantry frames can be raised to a stowed position inboard of the maximum beam.

The third method is to slide each gantry outboard from a stowed position within the maximum beam of the ship and to provide supporting structure over the gantry frames for the entire length of the two holds to be serviced.

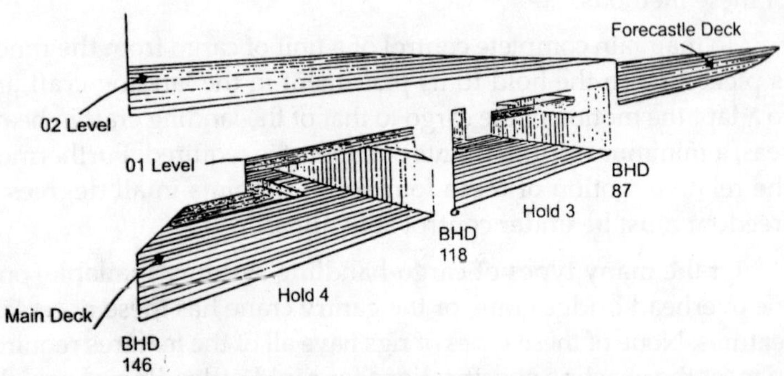

Fig. 8.1 : Main deck and structural arrangement for sliding gantry.

The after-starboard gantry is in the stowed position. The other three are in the lowered or extended position. Two LCM-8's are shown in stowed position on the starboard side of the main deck over hold 4. The starboard LCL is shown on the main deck over hold 3.

To provide an adequate base for the topping-lift attachment, structure and to provide space for the cargo-handling machinery, a centerline house is shown extending from bulkhead 146 forward to bulkhead 90. The house is 20 feet high from the main deck to the 01 level and is about 18 feet wide. The topping-lift attachment structure runs from the forward end of the deckhouse to bulkhead 90 and is supported by pairs of trussed king posts at bulkheads 90 and 118 and at frame 146 above the house top. The bridge structure between the king posts extends 70 feet above the main deck. In the topped position, the gantry frames extend 100 feet above the main deck.

Although the hinged-gantry system has some desirable features when compared with the sliding gantry, it also has many obvious disadvantages. The requirement for a continuous centerline deck-house prevents the offloading of landing craft or cargo from either side of the ship.

In order to use a sliding-gantry system, it is necessary to bridge over the two sets of landing craft stowed on the main deck. It would, of course, be possible to stow the landing craft on a higher deck and to provide an alternate king post and boom system for launching them. This would permit the bridging structure to span only the hatches. It would, however, involve a greater total vertical moment and inefficient use of the cargo-handling equipment.

Since the bridging structure is necessary, it is only logical to use it to maximum advantage in fulfilling other performance requirements of the ship. A fairly deep girder is needed to attain the necessary section modulus. If the upper flange of this girder is a continuous deck, it can serve several other purposes. Figure 10.2 shows such an arrangement.

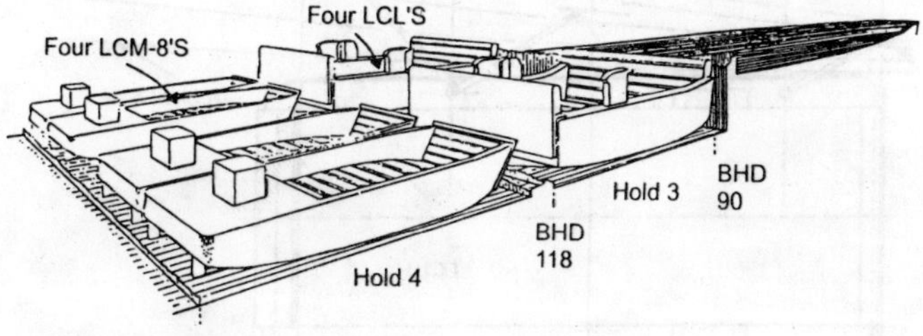

Fig. 8.2 : Main deck arrangement of stowed landing craft.

The upper deck shown here is at the 02 level or 29 feet above the main deck. It extends over the full beam of the ship from port to star-board and extends forward from bulkhead 146 to frame 84 where it then drops to the forecastle deck which is about 15 feet above the main deck.

No structure is required above this deck for cargo handling. It therefore provides a clear area for helicopter landing, 104 feet wide and 168 feet long. The deck provides complete weather cover for cargo-handling operations in holds 3 and 4 and, in combination with other elements of the bridging girder structure, it provides internal space for cargo-handling machinery and stowage.

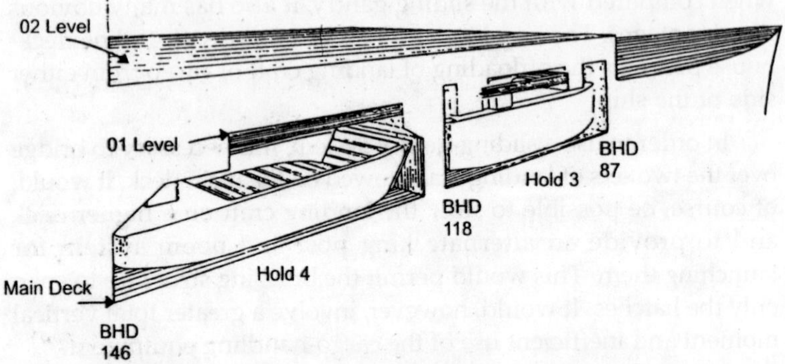

Fig. 8.3 : Outboard view of stowed craft and bridging structure.

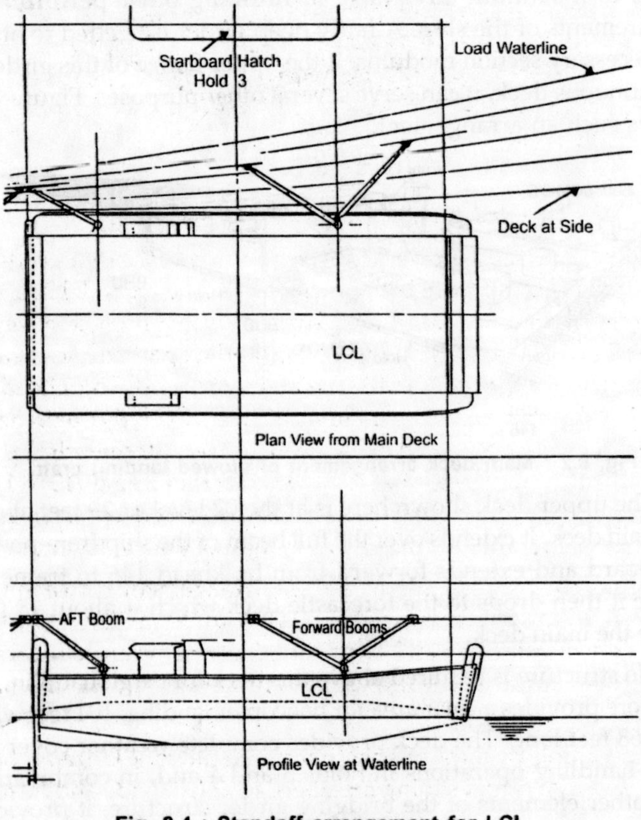

Fig. 8.4 : Standoff arrangement for LCL.

The primary girder over hold 4 is made up of the 01 level deck and a deck at the 02 level and extends forward from bulkhead 146 to bulkhead 118. It gives a girder length of 84 feet and a depth of 9 feet. At the after end of this girder, it is given additional support below by a 5-foot-deep transverse box girder between frames 139 and 143, which increases to a depth of 10 feet at bulkhead 146. At the forward end, the longitudinal bridging girder is supported by a transverse bulkhead between the main deck and the 01 level, 8 feet abaft bulkhead 118, and a transverse girder of trapezoidal shape which rises from a depth of 8 feet at this bulkhead to a depth of 5 feet at frame 123.

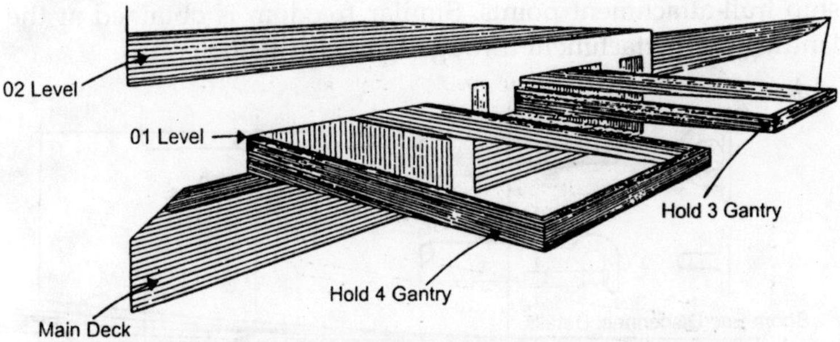

Fig. 8.5 : Starboard sliding gantry frames extended.

The primary girder over hold 3 is made up of the 01 level deck and a deck 7 feet below it, both of which extend from bulkhead 118 forward to frame 84. Additional support is given by transverse bulkheads which extend to 12 feet off centerline port and starboard at frames 90, 96, and 112 above the main deck and a bulkhead at frame 87 which spans the beam of the ship between the main deck and the 02 level. Longitudinal bulkheads, 12 feet off centerline port and starboard, run from frame 87 to 96 and from frame 112 to 118. Transverse box girders, 12 feet fore and aft by 7 feet deep, run across the entire beam of the ship forward and aft of the gantry frame.

Figure 10–3 shows the main-deck arrangement between the bulkhead at frame 87 and bulkhead 146 with the six landing craft in the stowed position and with the girder structure cut away. Figure 10–4 depicts the same arrangement with the girder structure in place.

The sliding portion of each gantry frame is a U-frame box girder. The individual sliding members are 2 feet wide and 50 feet long. The 38-foot portion, which extends outboard, tapers from a depth of 5 feet at the deck at side to a depth of 2½ feet at the outboard end. The 12-foot inboard portion is 5 feet deep. The sections tying the ends of the pairs of sliding members have widths of 2 feet and depths of 2½ feet. The fore-and-aft dimension of each U-frame is 50 feet. The lower members of these box girders are 15 feet above rhe main deck.

8.3 Details of Standoff Boom Connections

The booms are attached to the hull of the ship about 19 feet above the mean waterline. They are free to pivot to any angle at the ship hull-attachment points. Similar freedom is obtained at the landing craft attachment through ball-joint action.

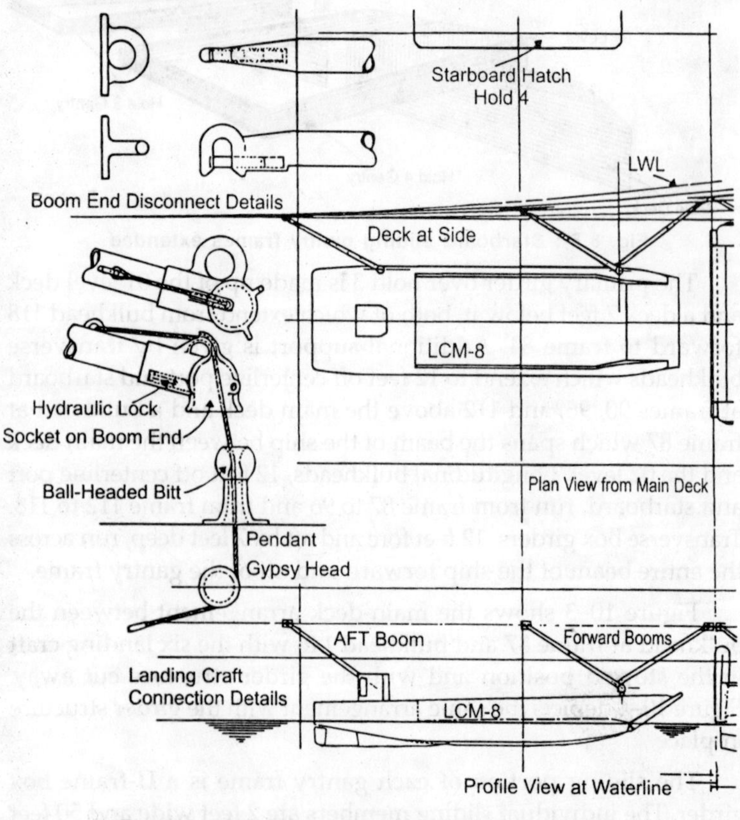

Fig. 8.6 : Details of standoff boom connections.

The booms are attached to the hull of the ship about 19 feet above the mean waterline. They are free to pivot, to any angle at the ship hull-attachment points. Similar freedom is obtained at the landing craft attachment through ball-joint action.

The telescoping of the booms is regulated by internal hydraulic pressure. The hydraulic lines are connected to an accumulator system within the FDLS. If the air pressure in the accumulator is adjusted, the telescoping spring stiffness of the booms can be varied. This boom arrangement allows freedom of motion of the landing craft in pitch, roll, and heave but restricts its motion in sidle, surge, and yaw to a degree dependent on the accumulator air pressure.

The accumulator air pressure has been reduced and the booms are fully extended. As will be discussed later, the lines leading from the ends of the booms to the extended gantry frame are also provided with a variable tensioning device. In the position shown, the line tension is at the minimum necessary to extend and support the booms. From the outboard ends of the booms, lines of relatively light rope, weighted at the ends, are suspended within reach of personnel on the landing craft.

Figure 10-8 shows details of the boom end attachments and plan and profile views of an LCM-8 alongside hold 4. As the landing craft comes alongside, the pendants are threaded into the ball-headed bitts and wrapped around air-driven gypsy heads on the landing craft. As the pendants are taken in by the gypsy heads, the sockets at the boom ends are pulled down on the ball-headed bitts where a hydraulically actuated pawl locks them in position.

The four lines leading to the gantry frame on the ship are then hooked to eyes located port and starboard at the forward and after ends of the landing craft cargo well. The accumulator air pressure, serving the booms and the line tensioning devices, is then increased until the boom telescoping stiffness and the line tension have reached their designed values.

When the ship is being made ready for offloading, prior to extension of the gantry frames, the lines from the outboard ends of the booms will be secured on the main deck at side, ready to be picked up and attached to the guidelines after the landing craft have been launched. In this condition the booms will be nested vertically along the side of the ship in the fully closed position.

This stowage may well be unsatisfactory for normal transit or standby operations. Therefore a quick release end fitting is shown so that the booms can be removed and stowed inboard. It is also assumed that quick-disconnect fittings for the hydraulic hose leading to the booms would be provided along the hull. An alternative might be to provide for the booms to retract directly into the hull to avoid the need for overside work in heavy weather.

The function of the telescoping booms in controlling surge, sidle, and yaw of the landing craft has already been discussed. The tension in the four vertical lines leading to the gantry frame should be made sufficient to impose a similar restriction on the pitch and roll of the landing craft.

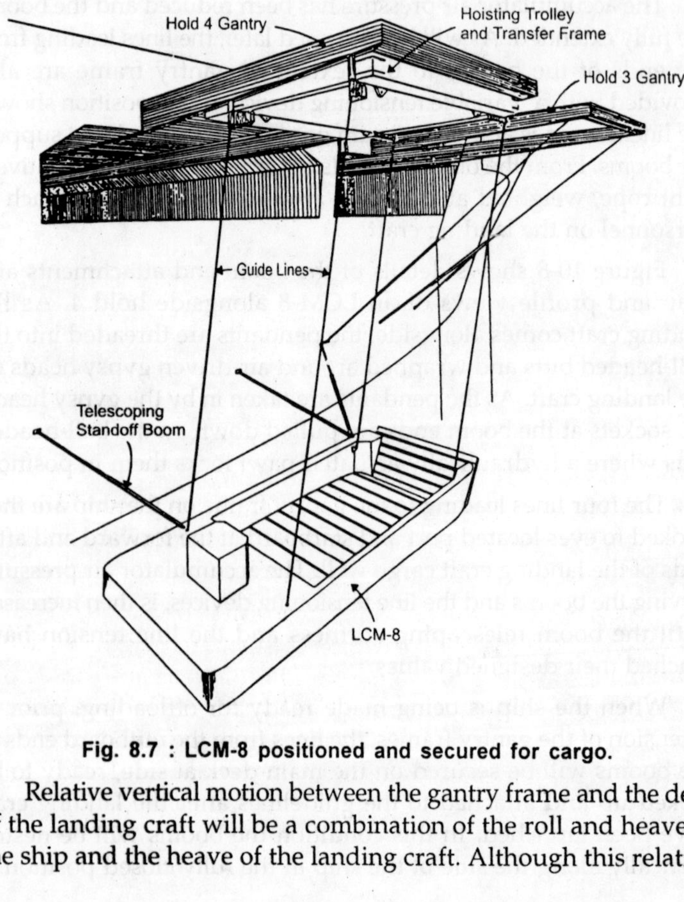

Fig. 8.7 : LCM-8 positioned and secured for cargo.

Relative vertical motion between the gantry frame and the deck of the landing craft will be a combination of the roll and heave of the ship and the heave of the landing craft. Although this relative motion may be reduced slightly in degree, it will still exist to a large extent in any appreciable seaway. The purpose of the four lines is not primarily to reduce this motion but to provide a measurement of its magnitude. They are used to guide the load from the gantry frame down into the landing craft and to alter the vertical motion of the load from ship motion to boat motion as will be described next.

8.4 The Handling System in Cargo–Transfer Frame

One basic element of the system is the cargo-transfer frame. This frame must be adaptable to the securing and supporting of any vehicular load, ranging from several jeeps to a 60-ton tank, and must remain attached to the vehicle from the pickup point in the hold to the point of placement in the landing craft. It must remain parallel to the ship decks from the time of attachment until the time it is being lowered overside. During the lowering operation, it must change its attitude to parallelism with the landing craft deck and must be given a vertical oscillatory motion equal to that of the landing craft.

The lateral dimensions of the cargo-transfer frame are dictated by the largest lateral dimensions of the vehicles to be offloaded. Its depth is dictated by the strength necessary to support the heaviest load. These specifications lead to the conclusion that the frame should have the length and width of a hatch opening and the depth of a hatch coaming. In effect, the four main deck hatch covers for holds 3 and 4 can also serve as cargo-transfer frames. These frames would have rounded corners and be 42 feet long and 17 feet wide, less than a few inches necessary for clearance in passing through the lower hatches. The hatch coaming depth of two feet would provide sufficient beam depth to support the maximum cargo unit to be lifted.

The frame can be a box girder consisting primarily of structural tees with the toe down and transverse intercostal plates installed to prevent racking. The top and sides of the frame can be plated over to provide water tightness and upper-flange girder strength. The slots between the toes of the tees on the bottom of the frame can provide a means of attaching slings for the vehicular cargo.

The following diagram indicates the attachment of the vechicle sling to the cargo-transfer frame.

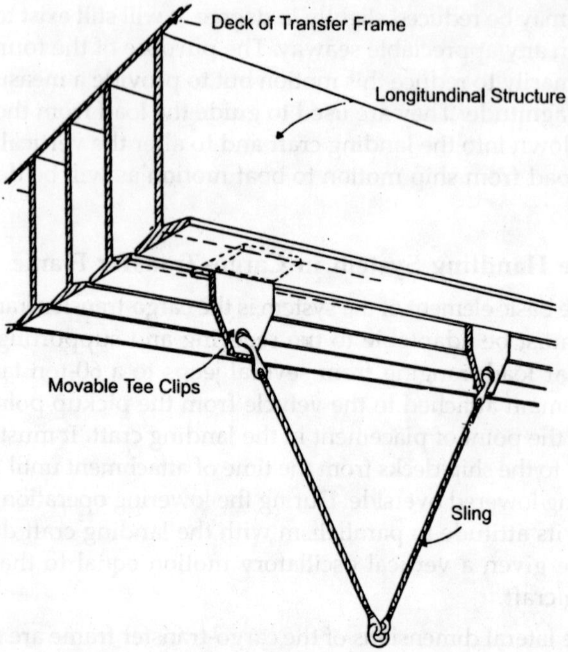

Fig. 8.8

Each vehicle can be fitted with a set of wire-rope slings which are arranged to fit between attachment points on the vehicle and the sliding tee eyes in the transfer frame. These slings can be designed so that the connection points on the transfer freame are symmetrically located relative to the horizontal position of the center of gravity of the vehicle. The length of the slings can be selected so that the top of the vehicle will be close to the bottom of the transfer frame. In this way any motion of the vehicle relative to the transfer frame can be restricted to an absolute minimum. (The slings for each vehicle would always be carried by that vehicle as a part of its equipment.)

One major advantage of using the transfer frame is in its capability of picking up and offloading a number of small vehicles for a single trip. Any number of vehicles which can be arranged within the 42-foot-by-17-foot hatch dismensions, and whose total weight does not exceed 60 tons, can be rigged with slings so that the vehicles fit in the same vertical dimension from tween deck hatch cover to transfer frame. Since the LCM-8 well deck dimensions are nearly the same as hatch dimensions, the transfer frame can carry a complete LCM-8 cargo load in one trip.

The depth of the hold from main deck to tank top is 72 feet. If the transfer frame is suspended at the tank top from a support 17 feet above the main deck, it would have a lateral displacement of 7 feet with only a 5' heel of the ship when in position to pick up a vehicle from the bottom of the hold. If the ship were rolling this small amount, it would be almost impossible to control the swinging cargo with this length of pendulum. Therefore, it is essential that the cargo frame be guided down through the hatches when hoisting and lowering.

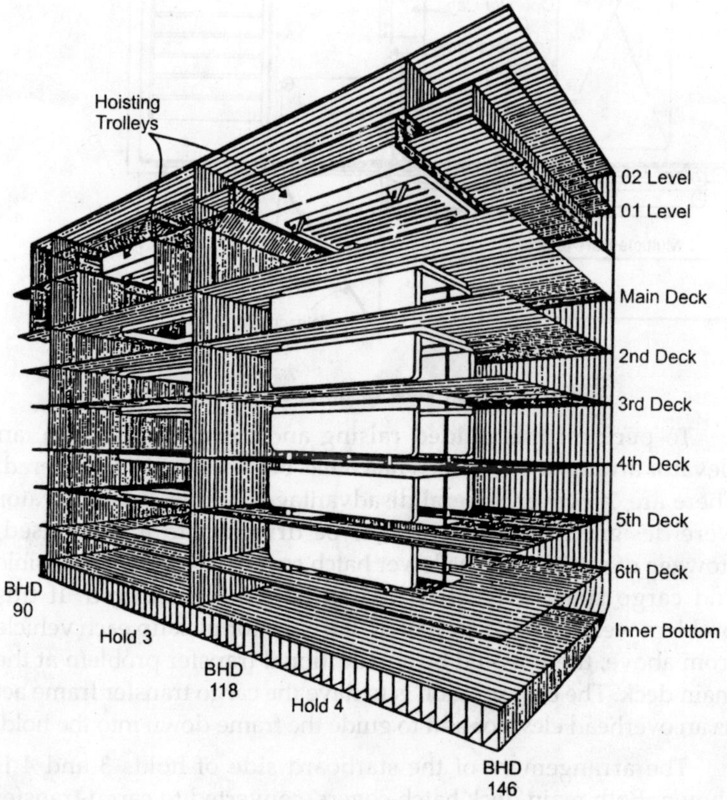

Fig. 8.9 : View of holds 3 and 4.

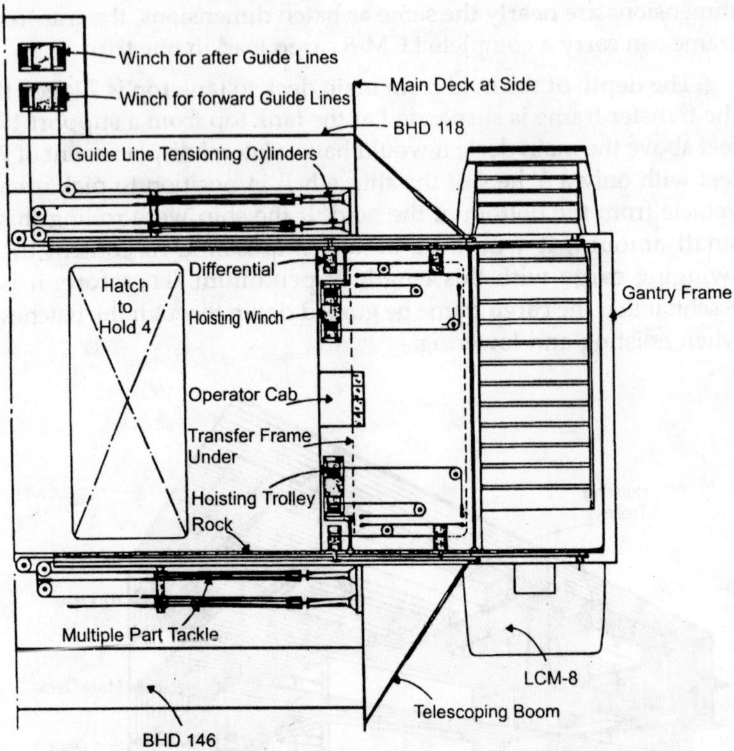

Fig. 8.10 : Plan view of offloading equipment

To perform this guided raising and lowering function, an elevator arrangement in each hatch has been seriously considered. There are, however, several disadvantages to this. If the elevator were designed for the vehicle to be driven onto it and raised, stowage of vehicles on the lower hatch covers would be impossible and cargo space utilization would be greatly reduced. If this problem were avoided by having the elevator pick up each vehicle from above, there would be an additional transfer problem at the main deck. The best solution is to have the cargo transfer frame act as an overhead elevator and to guide the frame down into the hold.

The arrangement of the starboard side of holds 3 and 4 is shown. Both main deck hatch covers, converted to cargo-transfer frames, have been hoisted to a nesting position in the hoisting trolleys. In hold 3 the tween deck hatch covers are shown in the closed position as they would be with the hold filled with stowed vehicles. In hold 4 the tween deck hatch covers are all shown in the open position as they would be when rigged for raising vehicles from the tank top.

The hatch covers are each in two sections, 42 feet long and 8.5 feet wide. They are hinged along the longitudinal hatch coamings so that when opened they form longitudinal bulkheads on each side of the hatch opening. The underside of each hatch cover section is fitted with toed–down channels recessed into the plating so that, when opened, vertical guide tracks are formed. Matching guide-track channel sections are built into the longitudinal hatch coamings to provide a continous pair of vertical tracks on each side of all open hatches.

On the bottom of the cargo-transfer frame, short extensions are fitted with spring-loaded rollers which match the guide-track spacing. As the transfer frame is lowered into the hold, the rollers fit into the guide tlacks, giving positive control of the frame down to the lowest position in the hold. The combination of secure attachment of the vehicle to the transfer frame and guidance of the transfer frame throughout the raising cycle provides positive control of the cargo to minimize swinging and prevent damage.

In the discussions which follow, regarding the hoisting system and motion-compensation system, we assume for simplicity that single wire lines and single sheaves and winch drums are used.

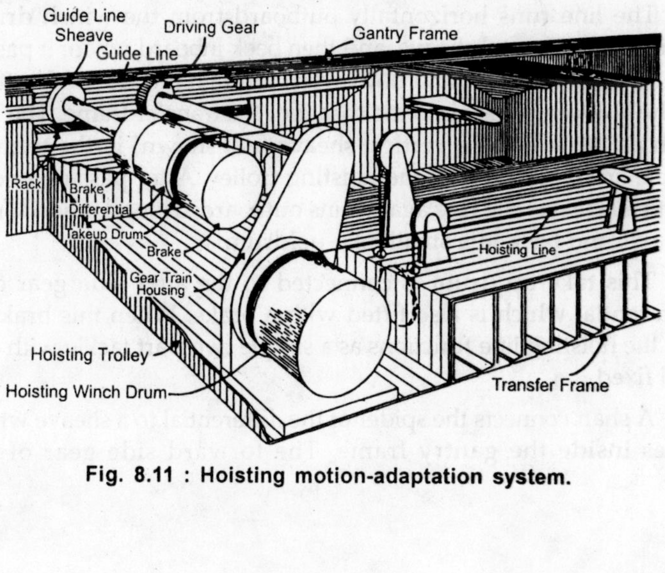

Fig. 8.11 : Hoisting motion-adaptation system.

Now let us try to understand more details about hoisting trolley and operator's cab.

The hoisting trolley is 46 feet long, 24 feet wide, and 5 feet deep. The transfer frame nests in the underside of the hoisting trolley so that the bottom of frame and trolley is a flush surface with a clearance of 15 feet above the main deck. The location of the transfer frame in the trolley is shown by the dashes.

The operator's cab on the inboard side of the hoisting trolley has controls for raising and lowering the transfer frame and for running the trolley in and out on the gantry frame. Windows provide a view of the operations below. (Matching transparent openings, not previously shown, will be required in the cargo-transfer frame.)

The hoisting trolley is fitted with two double-drum synchronized hoisting winches. Each drum stores the wire rope for lowering one corner of the transfer frame. With a clutch-brake and reduction-gear arrangement, each winch drives a spur gear which propels the hoisting trolley along a rack in the gantry frame. Another set of spur gears at the outboard end of the hoisting trolley are cross connected and ride along the gantry racks to give additional support to the trolley.

The hoisting lines from each winch drum in the hoisting trolley perform identical functions for the corner of the transfer frame which they support.

The line runs horizontally outboard from the winch drum, around a horizontal sheave, and then back inboard where it passes over a vertical sheave and downward through the recessed bottom of the hoisting trolley into the corner of the transfer frame. The line then passes around a vertical sheave (not shown) in the transfer frame and back up into the hoisting trolley. After passing over a vertical sheave, the line again runs outboard, around a horizontal sheave, and back to a small take-up drum.

This take-up drum is connected to the after side gear of a differential which is also fitted with a brake. When this brake is set, the hoisting line functions as a simple two-part tackle with one end fixed.

A shaft connects the spider of the differential to a sheave which rides inside the gantry frame. The forward side gear of the differential, which is also fitted with a brake, is free to rotate when the after brake is set.

If we shift now to the guideline attached to the forward port gunwale of the landing craft, the other element of the motion-adaptation system can be described. This guideline runs from its attachment point on the boat over a vertical sheave which is mounted on the inside of the gantry frame. It then runs inboard, makes a complete loop around the sheave connected to the differential spider, and around a horizontal sheave near the ship centerline. It then runs back out-board to the fixed end of a multiple-part tackle, out of the tackle, back inboard around a horizontal sheave, and then forward to a winch drum.

The outboard block of the multiple-part tackle is connected to the piston of a hydraulic cylinder. The hydraulic lines from this cylinder are connected to an accumulator. The cylinder is, in effect, a spring which can have its stiffness altered by varying the accumulator air pressure. When the winch drum is braked, the variation in guideline length due to the rise and fall of the boat is taken up in the multiple-part tackle due to the spring action of the hydraulic cylinder. A constant tension is also maintained on the guideline. The guidelines from the other three corners of the landing craft are fitted with similar constant-tensioning devices. Each drives the spider of one differential on the hoisting trolley.

Returning now to the forward inboard differential of the hoisting trolley, we see that the differential spider is in oscillatory rotation with a motion proportional to the movement of the forward port guideline around the connected sheave. The forward side gear of the differential will be rotating at twice this rotational speed.

As the hoisting trolley reaches the end of its outboard travel, the brakes on the driving spur gears are set and the winches lower the transfer frame. Since the hoisting winches are synchronized and one end of each hoisting line is fixed, the transfer frame will move downward in an attitude parallel to the hoisting trolley. Any vertical oscillatory motion will be the same as that of the gantry frame. In other words, at this initial lowering position, the transfer frame and its attached vehicles will have ship motion.

Just below the gantry frame roller guides are positioned around each guide line. As the transfer frame moves down from its nested position in the hoisting trolley, these roller guides slide into sockets in the transfer frame and lock in place. The transfer frame will then move laterally to follow the guidelines as it is lowered and thus will pick up any yaw, sidle, or surge of the landing craft as it moves downward.

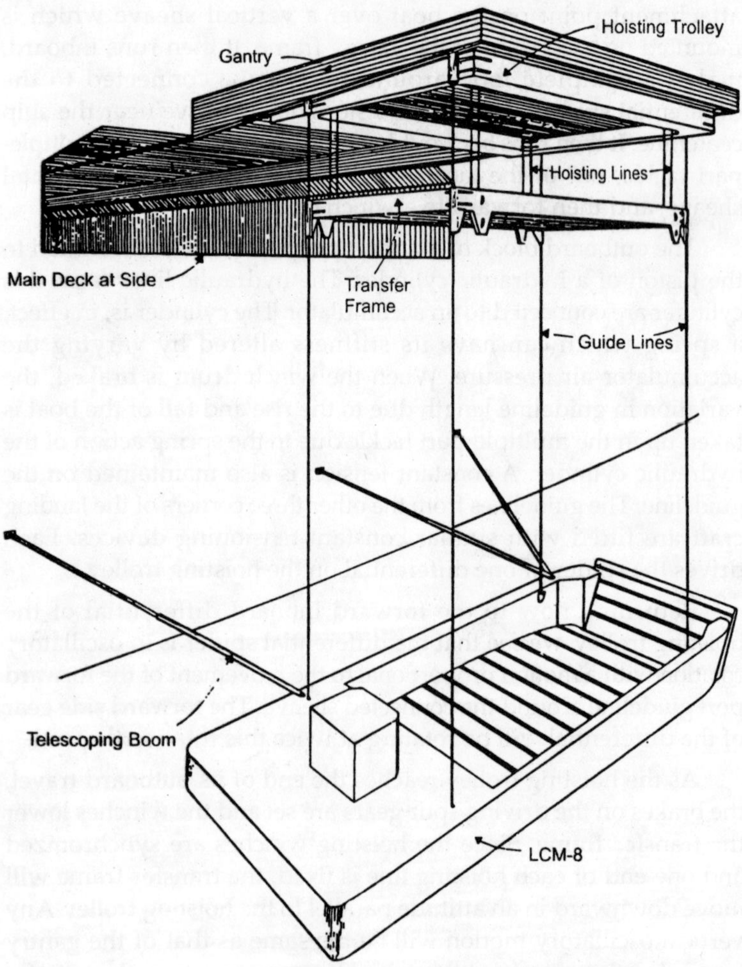

Fig. 8.12 : LCM-8 with transfer frame ready to lower.

When the transfer frame has been lowered to a position well clear of the gantry frame, the differential braking action is gradually shifted from one set of differential side gears to the other set. For the forward inboard hoisting line, the oscillatory rotation of the spider gradually transfers its driving action from the forward side gear to the after side gear. This in turn causes the hoisting line take-up drum to begin oscillatory rotation at twice the speed of the sheave around which the moving guideline is wrapped.

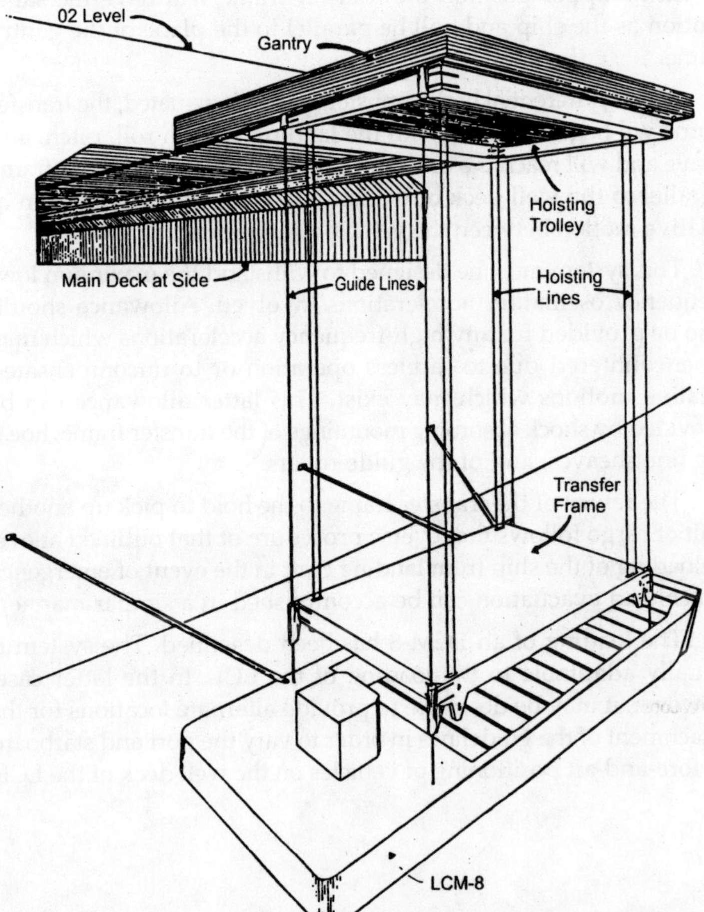

Fig. 8.13 : Transfer frame lowered to bottom position.

Thus the forward inboard corner of the transfer frame moves under the action of a two-part tackle of which one part is lengthening at a constant synchronized rate and the other part is varying in length at twice the rate of relative vertical motion of the forward port corner of the landing craft. The net effect is that this corner of the transfer frame will have the same vertical oscillatory motion as the boat and will close the distance between it and the boat at the synchronized lowering rate.

In its upper position the transfer frame will have the same motion as the ship and will be parallel to the plane of the gantry frame.

As the differential brake-transfer system is actuated, the transfer frame wil begin to move with the landing craft in roll, pitch, and heave and will reach the unloading position with the transfer frame parallel to the well deck of the landing craft with a minimum of relative motion between the vehicle and the craft.

This system must be designed to withstand the maximum low-frequency oscillatory accelerations involved. Allowance should also be provided for any high-frequency accelerations which may be encountered due to careless operation or to uncompensated relative motions which may exist. This latter allowance can be provided by shock absorbing mountings of the transfer frame, hoisting line sheaves, and of the guide rollers.

The return of the transfer frame to the hold to pick up another unit of cargo follows the reverse procedure of that outlined above. Reloading of the ship from landing craft in the event of emergency beachhead evacuation can be accomplished in a similar manner.

The loading of an LCM-8 has been described. The system is equally adaptable to the loading of the LCL. In the latter case, however, it may be desirable to provide alternate locations for the attachment of the guidelines in order to vary the port and starboard or fore-and-aft positioning of vehicles on the well deck of the LCL.

Chapter Nine
The Marine Science and Resources Policy

9.1 Introduction

Three important factors have stimulated the recent interest in improving the nation's technological effectiveness in the uses of the sea.

(1) the potential undersea threat to our national security ; (2) a recognition of the significance of science and technology as instruments of international affairs; and (3) an increased international awareness and interest in the potential benefits from exploiting marine resources.

The year 1966 marks a significant turning point in our national policy regarding exploration and development of marine resources. In that year, Congress passed two important acts: "The Marine Resources and Engineering Development Act" and the "National Sea Grant College and Program Act." The former created a National Council on Marine Resources and Engineering Development and an advisory Commission on Marine Science, Engineering and Resources–both charged with developing a comprehensive and coordinated national program to advance the marine sciences. The latter act was passed as an amendment to the former when it soon was recognized that attainment of many of the objectives of this "new thrust sea-ward" critically depended upon the availability of a group of skilled people to apply marine science and engineering techniques, equipment and research to developing the sea's resources.

The Marine Resources and Engineering Development Act of 1966 declared (it) "to be the policy of the United States to develop, encourage, and maintain a co–ordinated, comprehensive, and long-range national program in marine science for the benefit of mankind to assist in protection of health and property, enhancement of commerce, transportation, and national security, rehabilitation of our commercial fisheries, and increased utilization of these and other resources." Related to achievement of these national goals, this mandate identifies the following eight objectives:

- The accelerated development of the resources of the marine environment.
- The expansion of human knowledge of the marine environment.
- The encouragement of private investment enterprise in exploration, technological development, marine commerce, and economic utilization of the resources of the marine environment.
- The preservation of the role of the United States as a leader in marine science and resource development.
- The advancement of education and training in marine science.
- The development and improvement of the capabilities, performance, use, and efficiency of vehicles, equipment, and instruments for use in exploration, research, surveys, the recovery of resources, and the transmission of energy in the marine environment.
- The effective utilization of the scientific and engineering resources of the Nation, with close co-operation among all interested agencies, public and private, in order to avoid unnecessary duplication of effort, facilities, and equipment, or waste.
- The co-operation by the United States with other nations and groups of nations international organizations in marine science activities when such co-operation is in the national interest.

The National Sea Grant College and Program Act declared "that in view of the importance of achieving the earliest possible institution of significant national activities related to the development of marine resources, it is the purpose of this title to provide for the establishment of a program of sea grant colleges and education, training, and research in the field of marine science, engineering, and related disciplines."

Thus, the act has three purposes: (1) to increase and strengthen the trained manpower pool for participation in marine resources development; (2) to strengthen and improve applied research in the marine sciences, thereby increasing the knowledge and skill required to develop new techniques and equipment for effectively developing the marine resources; and (3) to strengthen and improve the process of information transfer (*e.g.*, marine advisory programs) in assisting persons or enterprises to further expand technology in marine resources development.

Total Federal Marine Science Program by Major Purpose[a]
(In millions of dollars)

	Estimated FY 1968	Estimated FY 1969	President's Budget FY 1970
International Co-operation and Collaboration	9.6	9.5	11.7
National Security	119.9	128.1	143.0
Fishery Development and Seafood Technology	40.1	43.8	44.5
Transportation	11.1	10.6	18.6
Development of the Coastal Zone	27.6	29.1	29.6
Health	5.3	5.4	5.3
Nonliving Resources	7.3	8.0	8.6
Oceanographic Research[b]	78.1	93.3[c]	93.5
Education	7.0	7.5	9.2
Environmental Observation and Prediction	28.8	31.6	32.1
Ocean Exploration, Mapping, Charting and Geodesy	75.7	83.0	99.1
General Purpose Ocean Engineering	19.2	18.3	29.5
National Data Centers	2.1	2.3	3.2
Total	431.8	471.5	528.0

[a] Many programs of the Departments of Defense, Commerce, Interior and Transportation and other agencies related to marine science affairs are not included.

[b] Research beneficial to more than one of the other major purpose areas.

[c] Includes $14.5 million for Coast Guard sub-polar oceanographic research ship. Funding for other ships is included in the relevant major purpose category.

Source: The third report of the President to the Congress on Marine Resources and Engineering Development, *Marine Science Affairs-A Year of Broadened Participation*, 11-12.

The Marine Resources Act requires that the President shall transmit an annual report to the Congress "which shall include a comprehensive description of the activities and the accomplishments of all agencies and departments of the United States in the field of marine science during the preceding fiscal year, and an evaluation of such activities and accomplishments in terms of the objectives set forth pursuant to this Act." The national policy was officially enunciated in the President's first report to the Congress:

The resources of the oceans can help us meet many of the challenges that face our Nation and the world today.

The vast food reserves of the sea must be developed to help end the tragic cycle of famine and despair.

The continuing pollution and erosion of our seashores, bays, estuaries, and Great Lakes must be arrested and reversed to safeguard the health of our people and to protect the resources of the sea.

The influence of oceans on the environment must be understood so that we may improve the long-term forecasting of storms, weather, and sea conditions; protect life.

Total Federal Marine Science Program by Department and Independent Agency[a] (In millions of dollars)

	Estimated FY 1968	Estimated FY 1969	President's Budget FY 1970
Department of Defense	240.6	262.3	297.9
Department of the Interior	70.5	76.2	78.2
National Science Foundation	38.1	35.0	45.4
Department of Commerce	33.6	33.4	49.0
Department of Transportation	15.4	34.1	32.3
Atomic Energy Commission	13.8	10.4	11.4
Department of Health, Education and Welfare	6.5	6.9	6.8
Department of State	6.6	7.0	7.1
Agency for International Development	3.0	2.5	4.6
Smithsonian Institution[b]	1.9	1.9	2.0
National Aeronautics and Space Administration	1.8	1.8	1.7
Total	431.8	471.5	528.0

[a] Many programs of the Departments of Defense, Commerce, Interior and Transportation and other agencies related to marine science affairs are not included.

[b] Excess Foreign Currency Funds are not included.

Source: *Marine Science Affairs-A Year of Broadened Participation*, 11–12.

9.2 The Fishery Resource Management

Fisheries science, as contrasted with fishery biology or other more limited concepts, is comprised of a number of segments. The most important of these are the resource, harvest, processing, marketing, legal, social and educational factors. These elements may be considered part of an overall system.

The study of the resource comprises investigations of the species, its sub-populations, the environmental parameters affecting them and the nature of the individual populations themselves, that is, their population dynamics.

The importance of knowledge of the environment and the species and its sub-populations, *i.e.*, systematic ichthyology, should be obvious. Such fundamental knowledge is basic to all other work. Inasmuch as the parameters affecting population dynamics can change from stock to stock, the nature of such an investigation involves a study of distribution and movement of the population through such techniques as tagging, tracking and examination of fundamental life history as well as through sophisticated numerical analysis of the variability of the species.

There are many factors which affect the distribution and abundance of the aquatic biota occupying a given geographical area. The most important factors appear to be the aquatic climate, particularly as related to temperature and salinity; the general nature of the habitat, that is, inshore-shelf inhabitants, benthic forms, pelagic life and distance from fundamental biotic source or "mother fauna."

Most fish groups are thought to have evolved in the inshore habitat, and with time the various species further differentiated and populated more specialized habitats as the pelagic and deep sea. In the course of evolution, the tropic areas evolved the greatest number of species per unit area, the temperate areas fewer and the polar waters the least. In terms of abundance, however, there appear to be fewer individuals of each species living in the tropic seas as compared with those living in waters of higher latitudes (or in water masses characteristic of those regions).

Numbers of individual species, that is, their abundance, is determined by habitat "space" within the biogeographic realm and available nutrients. Space as thus defined is a complex feed-back function of the kinds of animals and plants present and aquatic geography, while food available is a function of the aquatic climate and type of life at each trophic level.

The population dynamics of a single species of stock can be considered in the concept of inputs and outputs of a "black box." Biologically, four factors control the weight or biomass of a fish population.

Increases in populations are due to births or recruits and growth; the deletions are due to natural deaths or mortality, and fishing. Although much effort has gone into elucidating these processes, the black-box concept is not without meaning or unrealistic, as we have only rudimentary knowledge of the effect of density, food availability, and other dynamic features of the environment on a population and their effect on the total biomass. In its simplest form, the process may be considered as follows: from birth onward, the numbers of fish decrease due to natural mortality. From birth onward, the weight of an individual fish increases due to growth. The potential total weight of a population is the result of these two factors.

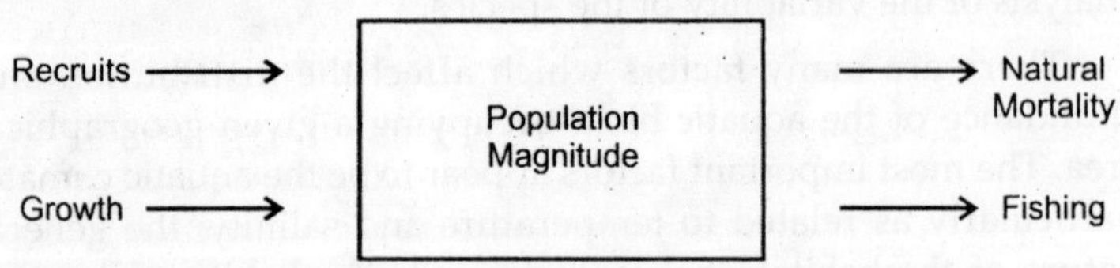

Fig. 9.1 : Factors controlling fish population biomass.

Examples of Regulatory Methods

Type of Regulation

Fishing Effort	Size at First Capture
Type and size of vessel	Mesh or hook size
Number of units of gear	Seasonal closures
Quotas	Fishing area limlts (to avoid nursery grounds)
Closure of fishing area	

Maximum biological yield may be achieved at a time of greatest population biomass, that is, at the point of greatest area under the two interacting rate curves. If yield is considered as some fixed proportion of the population at a given time, it is clear from Figure 8.2 that an early harvest would be the product of many numbers times low individual weight and would produce a low total yield (point A). Harvesting late (point C) also results in low total yield due to reduced numbers of individuals though of greater individual weight. At some intermediate point (point B), one would obtain the maximum yield.

The objective of fishery management in *biological* terms is a sustained harvest at optimum mean yield. These objectives are modified by economic requirements and sometimes by political or social values.

Management at the biological level attempts to influence the inflow and outflow from the population. One attempts to leave optimum numbers for spawning stocks by limiting catch, or through improving natural spawning areas or by creating artificial ones. Attempts to enhance yield through artificial propagation or improvement of the environment have also received attention. Natural deaths may be reduced conceivably through predator controls, but the only parameter through which man has any real control is fishing.

Many techniques for regulating fisheries have been employed and there are innumerable regulations in force.

9.3 Different Phases of Fishing

Fishing is essentially a hunting activity. As such, it comprises a strategy of three phases: search, identification and capture. Primitive as well as contemporary techniques of search comprise visual scouting, looking for such things as sea-bud activity, phosphorescent patches–brought about by fish activity at night, listening for sound produced by fish or locating particular environmental areas such as limiting temperatures where it is known that certain species are confined. Detection and location of fish schools can be carried out aerially, at the surface, or from beneath the surface.

Light-aircraft aerial reconnaissance for fishes has been employed for a number of years. Its success is largely dependent upon the depth of the species, as well as the depth of the water, the transparency of the water and the state of the air-water interface. Clearly, contrast is the significant factor in detectability. Occasionally, the species itself is never seen. Only the interaction between a surface-feeding school and the surface, causing ripples or otherwise a disturbed configuration, is visible. Because of the nature of the water surface, background scatter is such to make radar techniques ineffectual for location of fish populations. Recent tests by the Bureau of Commercial Fisheries in collaboration with NASA using high-flying aircraft provide evidence that dense concentrations or extensive shoals of fishes perhaps can be detected by environmental satellites. However, proceedings of recent remote-sensing conferences provide no "hard" evidence.

Infrared detection has some potentiality for the location of fish stocks. Such radiometric techniques measure surface temperatures and really define the environmental area or disturbed water occupied by the fishes.

They do not detect fish perse. The applicability of this technique on a broader scale is limited by the fact that radiometric techniques measure temperatures only at the water-air interface.

Penetration of light through the air-water interface is small except in the blue-green region of the electromagnetic spectrum. Laser offers a means of using extremely narrow band widths in the region of the best "window." Theoretical study implies that even under ideal situations, the use of laser beams would allow reconnaissance of the sea to depths of only 100 meters. To my knowledge, such tests have not been performed.

Obviously, any search, whether by air, surface or submarine, depends upon search patterns. The factors and constraints have been developed at length by operation research specialists in determining effective techniques for anti-submarine warfare. Such procedures are generally most useful in team operations and these are only little developed in fisheries as a whole.

Surface detection and location systems are essentially those techniques described for aerial use except that the sensor is brought down to the sea surface. Even here, every attempt is made to raise one's elevation. For example, the use of tethered balloons increases one's altitude over the deck. The only advantage of surface detection is that the variety of sensors is greater simply because of the nature of contemporary instrument development and the relative ease by which anomalies can be detected at the water-air interface.

The key to fish location and detection lies in the animal's habitat itself, that is, through the use of underwater detection systems. Sub-surface methods of detecting biological resources fall into two general categories: sound methods and non-sound methods (the latter being facetiously referred to as being "un-sound" by non-acoustic workers). Both methods can be employed in either an active or a passive sense. The technological development of sonar has been largely dependent upon military R & D. By its very nature, military sonar development tends to filter out the "noise" which is the very thing the fisherman seeks to identify and quantify. Consequently, in spite of the enormous developmental cost, it appears that sonar technology for biological purposes must be reengineered. Present systems permit detection of biological populations at ranges of no greater than 15 nautical miles. Pulse coding and doppler techniques in conjuction with side-scanning will allow some refinements, particularly in determining movements of fishes in terms of both direction and speed. Range of passive reception of noise produced by animals is dependent upon the frequency of the sound as well as its magnitude.

There is a real need in the sonic field for high resolution short-range equipment which will allow identification of targets with relative ease once a population is detected. This bears on the all-important problem of target identification. The range at which ider tification is possible is presently considerably less than the range of detection. Biological and physical data on the patterns of fish shoals in three dimensions would greatly facilitate the remote identification. Just as the field ornithologist can recognize, within a given number of types, the species of bird by the manner in which it flies, so fish schools also exemplify patterns in swimming and schooling which can be recognized. Remote techniques of identifying patterns are within the limits of contemporary technology. An unmanned active sonar interrogator, in either a mobile or fixed submarine buoy, could compare the pattern of a passing fish school with a library of likely species types stored in its computer memory. Identification of sound produced by fishes by the sonic signature is likewise feasible. Both of these developments require considerable engineering perfection before they can become practical realities.

The sonobuoy, developed during World War II for the location of submarines by aircraft has been employed for the location and tracking of fish schools. Essentially, it is a hydrophone attached to a radio transmitter with suitable receiving gear on board the aircraft. By employing several instruments simultaneously, direction and speed of an underwater target can be determined. Fishermen usually layout a pattern of such buoys and monitor them with existing radio equipment.

Another exciting potential for the future is the use of acoustic holography. Sonic holograms obtained from sonobuoys could materially revolutionize aerial or surface search and location techniques, as the 3D holograms would provide an estimate of the magnitude of the school as well as offer significant clues for its identification.

Non-sound methods of location and identification of biological resources are largely speculative. One possible visual means is the laser, particularly of the proper wave length. Scuba evidence to date suggests some promise for this technique. Even though it has been used at only very short ranges, the backscatter is much reduced over ordinary light. However, it is likely that shipboard range would be limited.

Perhaps the technique of most probable success would be the detection of resources on the basis of their odors. There is biological evidence for believing that fish can detect extremely small quantities of substances in their environment. Much more investigation is required, but if the basis for olfactory-sensory patterns can be detected, it would be it technically possible to develop recording spectrophotometers for tracing organic odors in the sea. With accrual of information on fish odor, it would be possible to identify fish schools by spectrophotometric means. An obvious ramification would be a dispersal of "a distasteful" odor to force or guide fish along a given path or the reverse technique to attract fish to a given point. There is considerable evidence that predators do make use of this principle to "home" in on odor and offal given off by fish schools.

The actual identification of any located stock must be either visual, behavioral or sonic. As previously implied, comparison of a monitored sense pattern with those in a computer memory is technically feasible and offers the ocean engineer an enormous horizon for development.

Once detected and identified, it is important to determine the magnitude of the resource. Interestingly enough, work in this area has progressed much farther than in identification. A number of techniques of interpreting echo returns from sonic devices into equivalent densities of fish stocks have been considered. One which has passed from the theoretical to the applied stage is that used in the United Kingdom (Cushing). The actual analysis is still largely manual but procedures for automation should be comparatively simple.

It will be obvious that search efficiency could be materially increased by combining the flexibility of aerial survey techniques with the range of underwater sensors available. It would be highly desirable to push for a breakthrough in such a procedure, as in areas where aircraft and reconnaissance vessels are currently employed in scouting for a combined fishing fleet, efficiency is very high. Such a procedure is presently in operation off Iceland. Possibly the future would see helicopters using "dipping sonar" for a sensor to obtain sonograms over large areas.

9.4 The Fishing Strategy

The second activity of our harvest system or fishing strategy is the capture operation. Many classifications have been proposed and are used for the several types of fishing gear employed in the capture of fish. One means suggests that all contemporary fishing gear can be divided into three categories:

Snares – fish caught directly through their own behavior

Lures – attraction of fish for their subsequent capture

Pursuit – fish caught by force such as surrounding nets and trawls.

The gill net is a good example of a snare; the long-line as used for tuna, with its bait or jig are examples of lures. The purse seine and trawl may be categorized as pursuit gear. Combinations of these techniques have been used, and of course other-classifications could be applied. There are several good reviews of various types of gear (Davis[2], Kristjonsson[4], Sundstrom).

It is apparent that most of these systems depend upon the nature of the environment and the inherent biological characteristics of the fish to provide concentrations or aggregations sufficiently dense to make capture worth the effort. Thus, contemporary gear is only useful for taking those types of fish that we are currently harvesting. Certainly there is room for expansion of present catches using such gear. Most of our fisheries are concentrated on peripheral sea areas or on the upper-most layers of the pelagic region. The potential harvest of deeper waters or mid-water layers has yet to be realized because of the incapability of the present gear to effectively fish these regions. If man is to fish other habitats such as the mid-open waters or take fishes "lower" in the food web, it will be necessary to devise more efficient straining systems, or develop methods which will cause fish to aggregate and therefore become more easily captured.

Most development of fishing gear and improvement to equipment used today has been an empirical accomplishment. Some of the laboratories have employed studies on visual acuity of fishes in attempts to make nets less visible, *i.e.*, less likely to be detected, as well as towing tank tests to measure and reduce drag of nets. Nevertheless, the analytical approach is missing. Very few data exist on the mathematical modeling of net efficiency or web straining capability. The varied problems offer a wealth of investigations for engineering talent.

There is a vast array of improvements that could be made to existing gear with suitable engineering development. The same statement can apply to ancilliary equipment. For example, improvement could be made in the materials themselves, both in strength and weight. There is a particular need for line gear of electrical conducting material so that the fish could be stunned upon capture, thus allowing for more rapid retrieval.

Even such apparently simple devices as line coilers are lacking on most fishing vessels. The development of a tapered cable of neutral or even positive buoyancy for fishing at intermediate depths would be a material advance to the fishery. There is a great need for communicating to the fisherman the precise position and location of the nets in relation to the fish. Because of the ship's roll, there is even an ambiguity of the amount of fish available on or near the bottom.

Improvement of otter boards or other hydrodynamic net-spreading devices would seem to be an obvious technological advance. The present nature of communication links leaves much to be desired and an accurate acoustic link between the capturing unit and control ship is still to be developed. The combination of lights, pumps and electricity has been explored only to a limited extent (Kristjonsson[4] and Vibert[7]). The effectiveness of such combinations has been demonstrated not only in the Russian kilka fishery but in recent work in the Gulf of Mexico in capturing shrimp in the daylight hours.

The wealth of technological problems involving fishing gear and equipment cannot be overemphasized. Why it has not attracted more engineering interest is beyond comprehension.

Another area directly related to the problems of the fisherman involves cost and weight of vessel construction. Whether fishing boats can be "generalized" and built on the module principle with standardized replacement parts remains to be determined. Much of the problem in the United States is due to the archaic law promulgated in 1792 to protect the U.S. ship building industry which requires a fisherman to use U.S. hulls in order to land fish at U.S. ports. Nevertheless, U.S. engineering technology certainly should be able to make use of the spin-off from other technological areas and employ light, strong materials for vessel and vessel machinery construction. For example, the ferro-cement means of hull manufacture looks most promising.

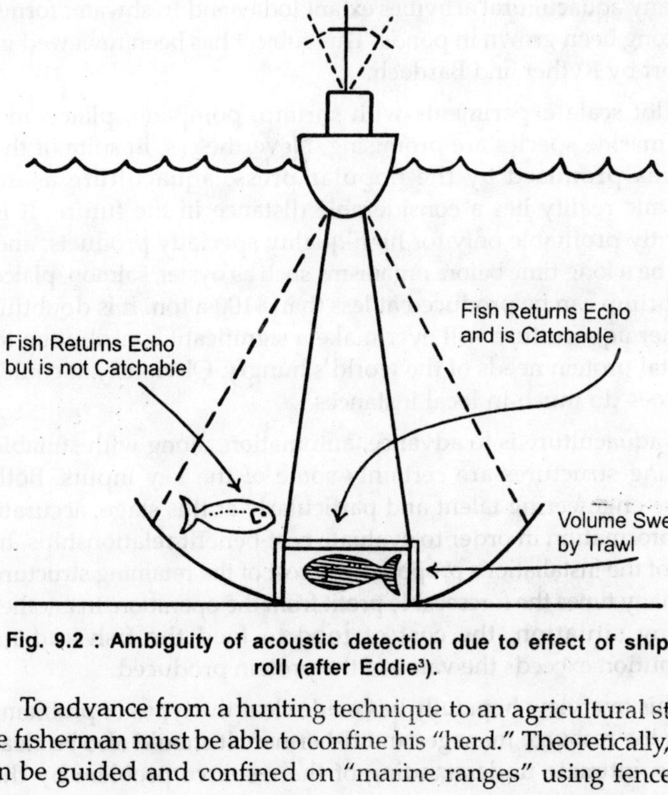

Fig. 9.2 : Ambiguity of acoustic detection due to effect of ship's roll (after Eddie[3]).

To advance from a hunting technique to an agricultural stage, the fisherman must be able to confine his "herd." Theoretically, fish can be guided and confined on "marine ranges" using fences or barriers of air bubbles or chemicals, ideally of the filter-bridge type that would let desirable species in and keep the undesirable out. One must be able to move either the fish to the food sources or vice versa. These management concepts are not beyond the reach of biology but require an enormous backup of engineering technology to become a reality.

For example, the development of a cheap, flexible and strong perforated hose whereby air bubbles or water containing a particular chemical could be used to produce a curtain or barrier would find use today in several inshore fisheries.

Another area which has received much attention and, unfortunately, much popular publicity is the artificial propagation of stocks or aquaculture. The development of such "animal husbandry" for fishes should certainly be encouraged in every way. Genetic manipulations of stocks could increase productivity. There are many aquacultural activities extant today and freshwater forms have long been grown in ponds. This subject has been reviewed in a report by Ryther and Bardach.

Pilot scale experiments with shrimp, pompano, plaice and other marine species are promising. Nevertheless, in spite of the panacea promised by the popular press, aquaculture as an economic reality lies a considerable distance in the future. It is currently profitable only for high-quality specialty products, and it will be a long time before organisms such as oyster, salmon, plaice and shrimp can be produced at less than $100 a ton. It is doubtful whether aquaculture will ever make a significant contribution to the total protein needs of the world's hungry. Obviously, it could, and does do much in local instances.

If aquaculture is to advance, automation, along with suitable retaining structures are certainly some of the key inputs. Both require engineering talent and particularly at this stage, accurate cost information in order to evaluate cost-benefit relationships. In some of the installations proposed, the cost of the retaining structure was many times the foreseeable profit from the operation. In another existing situation, the cost of food to feed the fish and its distribution exceeds the value of the protein produced.

This overview hopefully has made clear the types of problems facing the resource manager and fisherman. It should also be clear that an intimate understanding of the resource problem by the engineer is a necessity. Similarly, the resource manager-biologist must have a grasp of the technological capabilities. Until this forum and exchange of viewpoints can be accomplished, little progress is likely. Hopefully, with the development of national fisheries policies currently underway, there will be provision for input from the ocean engineering community. In addition to the direction and goals which should evolve from this science policy review, the author would urge consideration of the concept of critical mass. Fishery engineering and ocean engineering efforts are scattered. Very often one group has no knowledge of the other's activities. Individual efforts and funding are at low levels and progress proceeds slowly along most fronts. It would seem realistic and efficient to concentrate a critical mass of talent from all areas in a relatively few–or if necessary, single center–to meet the needs in this broad area. Some steps have been suggested by the U.S. Bureau of Commercial Fisheries in this direction but to no avail. These efforts are likely to continue coming to naught unless the pressure comes from the engineering community itself. Perhaps there is a method by which ocean engineers could communicate their interests to the National Academy of Engineering. Input from that level would necessarily receive consideration by policy makers.

CHAPTER TEN

THE OCEANOGRAPHIC VARIABLES

10.1 Classifications of Coastal Waters

Several classifications of coastal waters are possible. The continental shelf waters can be classified by the width of the continental shelf and the depth of the water on the shelf, i.e., 1) areas with narrow continental shelf and 2) areas with wide continental shelves, each of which can be subdivided into areas deeper than 50 meters and areas shall lower than 50 meters. A further division can be made by latitudinal zones; tropical, temperate and polar coastal water.

Most adjacent and semiclosed seas fall within the category of coastal waters. An oceanographic classification scheme of adjacent seas is shown in Figure 5-2. Again, further classification is possible: 1) semiclosed seas with unrestricted connection with the ocean, 2) semiclosed seas with restricted connection with the ocean, usually brackish, 3) semiclosed bays, and 4) estuarine environments. Large semi-closed seas such at the Mediterranean and some Indonesian seas are often excluded from the classification of coastal waters.

Some schemes have been used in the past for numerically determining the coastal influence by surface salinity. In many areas and conditions the coastal waters are separated from oceanic waters by sharp water mass discontinuities or fronts. The two water masses are differentiated through temperature, salinity, clarity and color of the surface layers.

Three dynamic processes have pronounced intensity and influence in coastal waters: (1) tides and tidal currents, (2) windstau effects and (3) bathymetrical and topographical modification of waves and currents. These processes promote turbulent mixing and thus affect the distribution of oceanographic parameters in shallow water.

A number of dynamic processes are peculiar to coastal waters. One of the more common is estuarian circulation in areas where fresh water runoff from the coast enters the ocean. The circulation is characterized by offshore flow of less saline water at the surface and shoreward flow of more saline water along the bottom. There is usually a stronger vertical stratification in coastal water near estuaries than in the open ocean, especially during the summer. During the winter the coastal waters in medium and high latitudes are usually isothermal and isohaline.

Another well-known dynamic process in coastal waters is wave refraction with its associated surf and currents. Longshore and rip currents return water which has been transported to the shore by the mass transport of waves.

The most pronounced dynamic process in shallow water is, however, the tides. Their nature and effects are quite varied from one location to another. Their analyses and prediction are the oldest subjects of environmental predictions in the oceans.

Table 10.1 : Continental Shelves of the World

Shelf	Area in 1000 km^2	Avg. depth or Depth range(m)	Avg. Benthos biomass g/m^2 (estimate)	Remarks (biological productivity, nature of bottom and coast, etc.)
Hudson				
Hudson proper	1010	60-150		Low productivity. Rocky coast.
Labrador	160	120		Low to medium productivity. Very rocky coast with numerous offshore islands.
West Greenland	180	75	100	Medium productivity. Very irregular and rocky coast with numerous fjords.
New Foundland-Nova Scotia				
New Foundland	400	120		High productivity. Rocky coast; many good harbours.
Nova Scotia	370	60-150		High productivity. Bottom very irregular with holes >200 m deep; many good harbours; coast rocky in many places.

contd.....

Table 10.1 contd...

Shelf	Area in 1000 km²	Avg. depth or Depth range(m)	Avg. Benthos biomass g/m² (estimate)	Remarks (biological productivity, nature of bottom and coast, etc.)
Iceland-Faroes	120	150	300	High productivity (except SE part of Iceland shelf). Rocky coast; many bays and fjords.
Spitzbergen-Barents				
Barents	550	70-150	100	Medium productivity. Low coast.
Spitzbergen	240	(120)		Medium to low productiv ity. Rocky and irregular coast.
NW European				
North Sea	570	35-(130)	200	N. part of the shelf is deep. Medium to high productivity. Low coast.
English Channel	90	35-(120)	40	SE part deep. Medium productivity. Many good harbours.
Irish Sea	380	80-150		Medium to low productiv ity. Rocky, irregular coast with many bays.
Norwegian	120	150		Medium to high productiv ity. Coast rocky with numerous fjords and offshore islands.
Baltic Sea	390	(100)	35	Low productivity. Fennonscandian coast rocky with numerous small offshore islands; SE coast low.
SW European				
Gulf of Biscay	80	130		Low productiviy.
West Pyrenean	50	(150)		Medium productivity. Narrow shelf with rocky coast.
Balearian and				
Gulf of Lion	40	50-100	15	Low productivity.
Adriatic	90	30-150 (70)		Low productivity. Eastern coast rocky, but has

contd.....

Table 10.1 contd...

Shelf	Area in 1000 km²	Avg. depth or Depth range(m)	Avg. Benthos biomass g/m² (estimate)	Remarks (biological productivity, nature of bottom and coast, etc.)
				many good harbors. Central, western and northern parts good for trawling.
Sicilian-Sidra				
Sicilian	30	50-100	10	Low productivity. The shelf is divided into western and eastern, parts. Few good harbours.
Sidra	100	60		Low productivity. Southwestern part relatively shallow.
USA-Atlantic				
New England	100	Coastal part 35 Offshore part 80		Medium to high productivity. Many protected bays with good harbors.
Carolina	120	25		Low to medium productivity. Shallow coast with offshore island and bars; Southern part coralline.
Bahama	130	15		Low productivity. Very shallow and reefy.
Gulf of Mexico				
Florida-Texas	450	40		Low to medium productivity. Outer part of Florida Shelf >120 m deep, shallow coast; many offshore bars.
Campeche	180	30		Low productivity. Shallow shelf, (except the northern corner); rocky and coralline; several offshore bars.
Carribbean				
Mosquito	110	25		Low productivity. Shallow shelf; rocky and coralline; many offshore bars & islands.

contd.....

Table 10.1 contd...

Shelf	Area in 1000 km^2	Avg. depth or Depth range(m)	Avg. Benthos biomass g/m^2 (estimate)	Remarks (biological productivity, nature of bottom and coast, etc.)
Venezuela	130	20-80		Eastern part has medium productivity, western low productivity. Rocky and coralline coast; irregular bottom.
Amazon	540	Southern part 40 N. coastal half 15-25 N. offshore half 70		Central and northern part along the coast has medium productivity; southern part and offshore area unproduc tive. Rocky and coraline, except off the Amazon river; no good harbor for medium of bigger boats.
Guinean	210	25	20	Low productivity. Many good sheltered places for harbors; central part rocky.
S. Brazilian	400	80		Central part has medium productivity; other parts low. Numerous lagoons and offshore bars; coast and N. part shallow (30m); N. part rocky and coralline.
Patagonian	1000	8-100		Medium to high productiv-ity. Southern coast and SE part rocky; Blanca and La Plata Bays shallow (*ca.* 20 m); few good harbors.
St. Helena	140	150	200	High productivity. Coast rocky; no good harbors.
Agulhas	110	100		Low productivity. Few good harbors; rocky coast.
Madagascar	180	15		Low productivity. Very coralline and rocky; no good harbors.

contd.....

Table 10.1 contd...

Shelf	Area in 1000 km^2	Avg. depth or Depth range(m)	Avg. Benthos biomass g/m^2 (estimate)	Remarks (biological productivity, nature of bottom and coast, etc.)
Island-shelves in Central Indian Ocean	240	40-80		Medium productivity. Most of the shelves are ocean coralline.
Zambesi	60	30		Low productivity. Generally rocky, except off Zambesi River
Persian Gulf	240	35		NW part has good bottom and medium productivity; SE part rocky and very shallow; N. part rocky and deeper with rocky coast.
Bombay	300	NE-20 Rest-80		NE part is shallow, has good trawling grounds and is productive. Central and S. parts are rocky, with medium production (except Wadge Bombs, which has productive trawling grounds)
Bengal-Burma				
Bengal	170	N-15 Rest-50-100		Northern part shallow and with medium productivity; remainder has low productivity.
Burma	380	(80)		Low productivity; Rocky coast with numerous offshore islands; coralline.
South China	2300	50		Low productivity (except Gulf of Siam). Shallow coastal area; bottom rocky & coralline.
Tong-King Hong-Kong	480	Coastal half 35 offshore half 85		Low productivity; Partly rocky; N part has numerous small offshore islands.
East China Yellow Sea	1000	W part 30 E part 80	175	Medium productivity E. and S. coast rocky; shallow and mueddy off deltas.

contd.....

Table 10.1 contd...

Shelf	Area in 1000 km^2	Avg. depth or Depth range(m)	Avg. Benthos biomass g/m^2 (estimate)	Remarks (biological productivity, nature of bottom and coast, etc.)
Okhotsk	580	140	220	Medium productivity. Kamchatka and Sakhalin coast more shallow (60 m); rocky coast; few good harbors.
Arafura	1350	(70)		Low productivity. Rocky and coralline; no good sheltes along the coast.
NW Australian	300	80		Low to medium productivity. Irregular bottom; NE coast has many offshore islands; other parts of coast do not have good shelters.
S. Australian	350	40-130		Medium productivity. Only central part has some good shelters.
Tasmanian	170	60-80		Medium to high productiv ity. Partly rocky.
New Zealand	250	70		High productivity (except N. part) rocky coast.
Queensland	220	<40		Low productivity. Extremely rocky & coralline.
Bering Sea	1200	NE-35		Medium to high productivity. (NW part low productivity). Many great sheltered bays; coastal area in some places rocky; low bottom water temperatures.
North Siberian	2600	W-25 E-40	25	Low productivity. Low coast, mostly ice covered.
Black Sea			(300 Azov Sea)	Sea of Azov very productive. (Depth 15 m). Relatively large shelf in northern part. Depth of NW shelf *ca.* 50 m.

10.2 The Intensity of Sea-Air Interactions

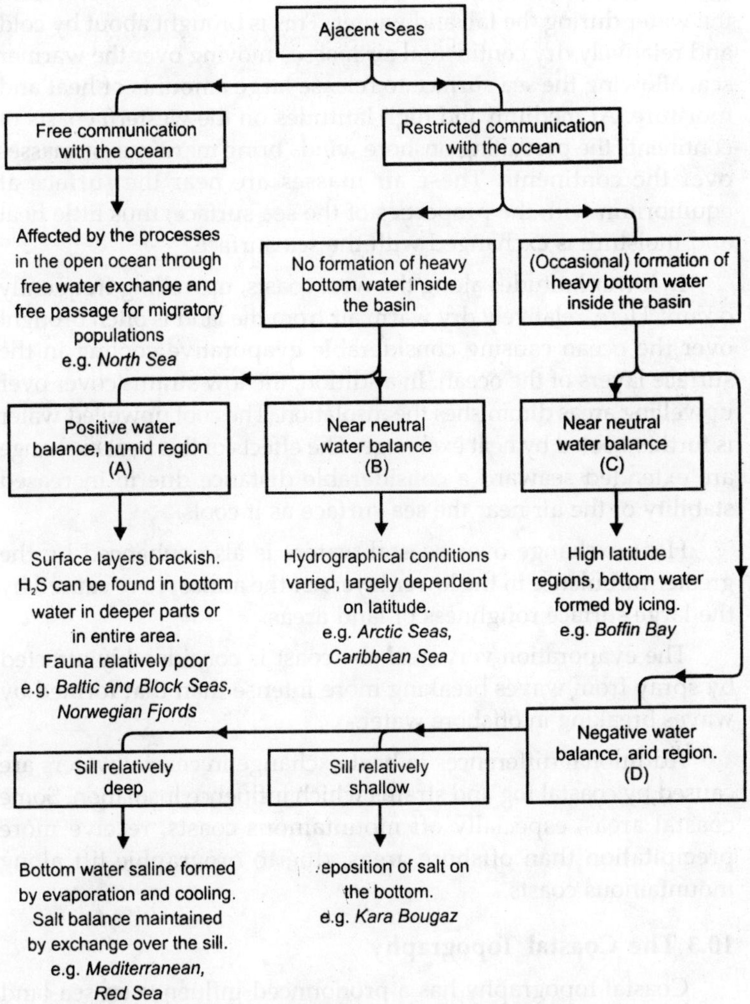

Fig. 10.1 : Oceanographic classification scheme of the adjacent seas (after Hela and Laevastu[2]).

The intensity of sea-air interactions with respect to heat exchange is greatly determined by the difference of the air and sea surface properties in coastal waters. Great differences exist between the east and west coasts of the continents in this interaction because of the continental influence. The medium and high latitudes off the eastern coasts of continents are characterized by great heat loss from the water during the fall and winter. This is brought about by cold and relatively dry continental air masses moving over the warmer sea, allowing the sea surface to release large amounts of heat and moisture. At medium and high latitudes on the western coasts of continents the prevailing onshore winds bring maritime air masses over the continents. These air masses are near the surface at equilibrium with the properties of the sea surface; thus little heat and moisture is exchanged with the sea surface.

In lower latitudes along the west coasts, upwelling frequently occurs. Here, relatively dry warm air from the land is often brought over the ocean causing considerable evaporative cooling in the surface layers of the ocean. In addition, the low stratus cover over upwelling areas diminishes the insolation. The cool upwelled water is further cooled by heat exchange. The effects of this heat exchange are extended seaward a considerable distance due to increased stability of the air near the sea surface as it cools.

Heat exchange over coastal waters is also enhanced by the greater turbulence in the lower layers of the atmosphere caused by the large surface roughness of land areas.

The evaporation very near the coast is considerably affected by spray from waves breaking more intense than that formed by waves breaking in offshore water.

Additional differences in heat exchange in coastal waters are caused by coastal fog and stratus which influence insolation. Some coastal areas, especially off mountainous coasts, receive more precipitation than offshore areas, due to orographic lift along mountainous coasts.

10.3 The Coastal Topography

Coastal topography has a pronounced influence on sea-land interactions, especially along mountainous coasts. The coastal topography determines the sheltering and fetch effects which limit the generation of waves. Local coastal winds such as the mistral, tramontana, Santa-Anna, etc. are a function of coastal topography.

An integral part of the sea-land interactions is the sea and land breeze caused by diurnal heating and cooling at the land. It occurs with great regularity on many medium and low latitude coasts but at higher latitudes only during warm seasons.

The most significant oceanographic sea-land interactions manifest themselves in the form of runoff from rivers and streams. This runoff has considerable influence on the oceanographic conditions near the coast, creating special environments in estuaries and in semiclosed bays. The runoff changes the salinity, turbidity and to a lesser degree the temperature of the coastal waters. Thus, the salinity anomalies tend to be more local phenomena than the temperature anomalies in the ocean.

Another major characteristic of coastal waters is the high basic organic production. The proximity of land provides an abundance of nutrients, both by runoff and by wind transport.

Ice formation in coastal waters depends largely on the continentality of the coastal climate, the salinity stratification, water depth and the exchange of water between the offshore and coastal area.

Table 10.2 : Comparison of Coastal and Oceanic Environments

Parameter and/or Process	Oceanic	Coastal
Temperature	Relatively uniform with moderate to low ambient thermal noise (except near major current boundaries). Vertical and horizontal gradients moderate to low. Changes determined mainly by heat exchange and advection.	Moderate to high ambient thermal noise, greatly influenced by coastal topography, bathymetry and upwelling-anstau. Vertical and horizontal gradients moderate to high. Changes determined by heat exchange, advection; and mixing, especially by tidal current mixing. Horizontal gradients towards coast usually positive in summer, negative in winter.
Salinity	Uniformly high with small horizontal and vertical gradients. Determined mainly by evaporation-precipitation balance.	Relatively low, especially in bays and near estuaries. Horizontal and vertical gradients moderate to high. Determined by runoff and evaporation-precipitation balance.

contd.....

Table 10.2 contd....

Parameter and/or Process	Oceanic	Coastal
Permanent Currents	Predominantly thermohaline, speed relatively uniform and low.	Thermohaline component greatly modified by local bathymetry. Considerable accelerations on continental slopes, due to anstau.
Wind Currents	Determined by average prevailing wind fields.	Wind currents greatly modified by coastal topogra phy, anstau and upwelling.
Tidal Currents	Usually weak	Usually dominates other current components.
Waves	Determined by local wind fields and by swell propagation.	Modified in shallow water due to bottom influence. Generation affected by sheltering effect of land (limited fetch).
Sea-air exchange process	Determined by surface meteorological systems and processes; patterns correspond to the dimensions of meterorological systems.	Exchange processes can be intense at times due to large differences in tempeature and moisture as continental air masses move over the sea surface.
Sea level	Variations usually small, dependent on tides and atmospheric pressure.	Variations usually large, caused by astronomical tides, co-oscillations of bays, atmospheric pressure and wind-induced anstau.
Turbidity (and visibility in water)	Low turbidity; largely determined by standing crop of phytoplankton	High turbidity due to suspended matter, transported by runoff and stirred up from bottom by wave action.
Biological populations	Plankton population relatively uniformly distributed. Productivity medium to low (Except near current boundaries). Standing crop of fish relatively low and dispersed	Plankton population relatively patchy. Productivity generally medium to high. Standing crop of fish high, many species occurring in schools. Dirunal and seasonal migrations pronounced.
Pollution	Low	Can be considerable along populated shores and near estuaries.

10.4 Food from the Sea

Food experts keep reminding people that the rate of increase of the food supply is barely keeping pace with world population growth-moreover, that one-half of the world's population-1.5 billion people-is afflicted by protein malnutrition. In view of these unsettling statistics, food from the sea offers the high promise of meeting a part of this worldwide need.

The current U.S. policy with regard to marine resources is aimed at both rehabilitating the fishing industry into a viable and efficient sector of the economy and accelerating the contribution of marine resources to the continued expansion of the economy. The policy is also aimed at encouraging international cooperation in exploiting the full potential of the seas through application of modern science and technology and economic assistance. To this end, the United States is embarked on a long-range program to exploit the food potential of the oceans to help feed the undernourished people of the world. This is in accord with the broad strategy of the War on Hunger Office of the Agency for International Development. A food-from-the-sea service was organized within the agency to assist in the technological

Table 10.3 : Total Federal Marine Science Program by Function Summary (In millions of dollars)

	Estimated FY 1968	Estimated FY 1969	President's Budget FY 1970
Research and Development	245.4	248.3	297.9
Research (basic and applied)	116.4	126.7	150.6
Development of new equipment and technology	129.0	121.6	147.3
Investment	55.5	77.8	70.1
Ships	7.6	15.1	2.3
Major equipment	29.3	42.4	47.5
Shore facilities	13.9	15.5	13.3
Other	4.7	4.8	7.0
Operations	130.9	145.4	160.0
Surveys	106.0	118.7	130.9
Services	18.4	19.5	21.4
Other	6.5	7.2	7.8
Total	431.8	471.5	528.0

Source: Marine Science Affairs-A Year of Broadened Participation

development and distribution of FPC (fish protein concentrate) a tasteless, odorless protein concentrate which can be combined with flour, cereals and other processed foods. Pilot production operations have shown the feasibility of producing ten grams of FPC-supplying the minimum daily animal protein nutritional requirement-at less than one cent.

Estimates for increasing the world fisheries catch with current techniques range anywhere from a doubling of the present output of approximately 60 million metric tons to possibly five or more times. Although there are vast amounts of potentially exploitable resources off the coasts of North America, relatively few species are actually caught by U.S. fishermen. Since 1956 the United States has dropped from second to sixth place among the world's fishing nations. What are some of the factors cited as reasons for this relative decline? Basically, they are subsumed under what may be termed "institutional constraints." That is, laws, customs, and economic regulations-national and international-which limit the fish-catching ability and capacity of fishermen.

World Catch[a]/of Fish, Shellfish, Etc. by Leading Countries 1954-66
(In billions of pounds)

Country Ranking by 1966 Catch Figures	1954	1956	1958	1960	1962	1964	1966[b]
Peru	0.4	0.7	2.2	7.9	15.3	20.2	19.4
Japan	10.0	10.5	12.1	13.7	15.1	14.0	15.6
China	5.1	5.8	8.9	12.8	NA	NA	NA
USSR	5.0	5.8	5.8	6.7	8.0	9.9	11.8
Norway	4.5	4.8	3.2	3.4	2.9	3.5	6.3
United States	6.1	6.5	5.9	6.2	6.6	5.8	55.1
World	60.5	67.1	72.4	87.1	102.0	114.0	125.0

[a]/Total commercial catches, including fresh water fish, on a live weight basis. Weights of shells of univalve and bivalve mollusks are included.

[b]/Preliminary.

NA = Not available.

Source: The second report of the President to the Congress on Marine Resources and Engineering Development, *Marine Science Affairs-A Year of Plans and Progress.*

These might include restrictions on fishing area, season, or types of vessels, gear and techniques allowed.

What rationale is there for imposition of these impediments on the efficiency and effectiveness of fishermen? Some of the reasons given are: (1) to maintain a sustainable fish catch; (2) to rconcile conflicting user interests, *e.g.*, between sport and commercial fishermen; (3) to insure use of American-built fishing vessels; or (4) to insure that small, individual fishermen are not placed at a competitive disadvantage compared with large fishing organizations. These policies have resulted in serious economic inefficiencies imposed on American fishing activities. They have led to the fragmentation of the industry into small units, finding it difficult to accumulate capital in sufficient amounts to adopt the latest equipment and techniques. The only real distant-water operations of the U.S. fishing fleet are the tuna and shrimp fisheries.

This approach to the U.S. fisheries policy has conditioned the strategy of the fishing industry to continually seek governmental assistance through subsidies and protection from foreign competition, rather than adoption of more efficient fishing techniques and locating new resources. Federal government assistance to the fishing industry tends to reflect this kind of strategy.

A breakdown of the U.S. Bureau of Commercial Fisheries budget for programs to foster the fisheries' growth and diversification shows the following allocation of expenditures: 63% devoted primarily for gathering data on abundance and distribution of fish stocks and attempting to estimate maximum sustainable yields; 20% applied to research into processing and marketing; while only 10% is directed to development of new fishing technology for greater catching efficiency or locating techniques, or new strategies of fishing.[9] It is generally felt by critics of U.S. fisheries policy that these constraints, resulting in limited economic viability of the fishing industry, have hampered the effective adoption and application of recent advances in ocean science and technology.[10]

Recently accelerated advances in marine technology enable countries to exploit ocean resources farther away from home and in waters deeper and previously inaccessible. Recognizing both the advantage of international cooperation in the scientific study of the sea and potential conflicts between competing uses of the sea, there is a growing consensus that a multi-national approach is needed to ensure the peaceful uses of the sea. The traditional "freedom of the seas" doctrine of international law, for example, can actually lead to intense rivalry over marine resources and the uses of the seas. Unlimited entry of new and potential exploiters will generally prevent any sort of voluntary restraints on exploitation and utilization of resources by present participants. Actually, the pressure will be to exploit as fast as possible!

Almost all participants in the current debate on the law of the sea, therefore, generally agree that some sort of inernational legal regime needs to be formulated which tends to limit entry and, at the same time, maximizes the net economic benefits from exploitation of marine resources so as to prevent misuse, overuse, destruction, waste, depletion and utter chaos. Two basic premises underlie this new thinking and approach to an effective international organization for controlling access to the ocean resources: (1) given the search for new food resources and the growing demand for energy and mineral resources, the marine technology which permits exploration and exploitation of marine resources is rapidly becoming available; (2) any major development of deep sea marine resources will require large capital investments and this in turn will require protection of access rights for continual, rational development and exploitation. With regard to these matters, two paramount issues are currently receiving considerable attention by the maritime nations of the world: What should be the seaward limit of the continental shelf? What resources are there beyond the continental shelf and who should control them?

Existing international law of the sea distinguishes between the internal waters and territorial sea of a nation; the high seas; the contiguous zone; the continental shelf; and the bed and subsoil of the deep seas, or the high seas beyond the continental shelf (see figure 3-1).[11] While these global areas have no precise geographic references, it is important to distinguish between the rights of the United States *vis-a-vis* all other nations in these areas, and the division of authority between the federal government *vis-a-vis* the coastal states in areas acknowledged to come under the nation's jurisdiction.

The Convention on the Territorial Sea and the Contiguous Zone states that within a zone of the high seas, which may not extend beyond 12 miles from the baseline from which the breadth of the territorial sea is measured, each coastal nation may exercise the control necessary to prevent infringement of its customs, fiscal, immigration, or sanitary regulations and may penalize any such infringement committed within its territory or territorial sea. Although the convention seemingly does not extend national jurisdiction, *in toto,* in the contiguous zone, coastal nations have claimed permanent, exclusive access to living resources up to 12 miles from the coast, including the territorial sea. Thus, the United States prohibits foreign vessels from fishing within its I2-mile "exclusive fisheries zone" except under a bilateral agreement.

The International Convention on the Continental Shelf gives the coastal nation the sovereign and exclusive right to explore the shelf and exploit its natural resources. If the coastal nation does not itself engage in these activities, or make express claim to the continental shelf, no other party may do so without its permission.

The continental shelf as defined by the convention includes "the seabed and subsoil of the submarine areas adjacent to the coast but outside the area of the territorial sea, to a depth of 200 meters (656 feet) or, beyond that limit, to where the depth of the superjacent waters admits of the exploitation of the natural resources of the said areas" and "the seabed and subsoil of similar submarine areas adjacent to the coasts of islands." The natural resources to which the coastal nation is given permanent, exclusive access include "the mineral and other nonliving resources of the sea bed and the subsoil together with living organisms... which, at the harvestable stage either are immobile on or under the seabed or are unable to move except in constant physical contact with the seabed or subsoil."

CHAPTER ELEVEN
ICE PROBLEMS IN NAVIGATION

11.1 Introduction

Navigation in temperate and arctic zones has always been affected by problems associated with ice conditions.

In the coastal waters of the North American continent above the American-Canadian borderline, ice conditions hamper navigation during a considerable part of the year.

The first example of an important navigation route affected by ice conditions is the St. Lawrence Seaway. The St. Lawrence Seaway was discovered and explored by European adventurers looking for a northwest passage. In 1535, Jacques Cartier sailed about 1000 miles into the interior until his ship was halted by rapids. The history of this waterway's development is a very long one-the last sections were constructed only in recent years.

One of the seaway's most essential problems is ice formation which makes the route unnavigable during 3-4 months of the year. Needless to say, this is quite an economic drawback to a sound seaway project development. In January, temperatures over the estuary average below 10°F.

Icebreakers have been used extensively on the seaway to break up ice jams to prevent serious flooding (1) and to assist ships trapped in the ice in early winter. More and bigger icebreakers, aerial ice surveys and improved aid to navigation are easing the situation.

The purpose of this chapter is not to give a complete account of all problems in this field but rather to indicate the existing problems for which solutions need to be found. We will now take a closer look at the problems associated with the attempts to find a navigable northwest passage as mentioned before and will discuss the recent attempt to find a modern solution.

The discovery of new and presumably rich oil fields along the north shores of Alaska (Prudhoe Bay) created the problem of transporting the oil to the industrial sites of the U.S. eastern seaboard. After carefully evaluating a number of alternatives, the oil companies involved (Humble Oil and Atlantic Richfield) decided that the problem could possibly be solved by using supertanker-icebreakers through the Northwest Passage. They decided to refit one of the largest tankers in existence and to make a voyage from the east coast of North America to Alaska and return; this experimental program cost approximately 40 million dollars.

The *Manhattan*, the largest tanker flying the U.S. flag, was chosen for this adventure. The *Manhattan* had a 108,400 dwt capacity, measured 940 ½ feet long with maximum draft of 52 feet, 9 3/8 inches and had twin screws of 43,000 shp in total.

Before we go into the technical details of this operation, it is interesting to remember some of the historical attempts at finding the Northwest Passage. The search for such a passage from western Europe to the Far East encompasses a dramatic story that began almost five centuries ago.

In 1497, a few years after Columbus came back to Europe from his trip to the West, the English sent a Genovese navigator named Giovanni Cabato, better known as John Cabot, to seek the northwest connection between the Atlantic and Pacific oceans. This unsuccessful attempt was repeated several times in later years.

In history books we read the names of Martin Frobisher (1576), John Davis (three attempts between 1585 and 1587) and Henry Hudson (1610); Davis reached Lancaster Sound, the entrance of the true passage, but his discovery was not recognized at that time. Henry Hudson, with his ship, the *Discovery*, sailed into the bay named after him; he found death when his crew put him adrift in a boat to perish in the icy winter.

In 1616 the *Discovery*, under the command of Robert Bylot, a rehabilitated mutineer, reached Smith Sound at a latitude of 78°, which was not reached again for 237 years. In 1819 Parry sailed through Lancaster Sound and reached 113° west longitude. The two ships of Sir John Franklin disappeared forever in 1845 after his expedition had been sighted entering Lancaster Sound.

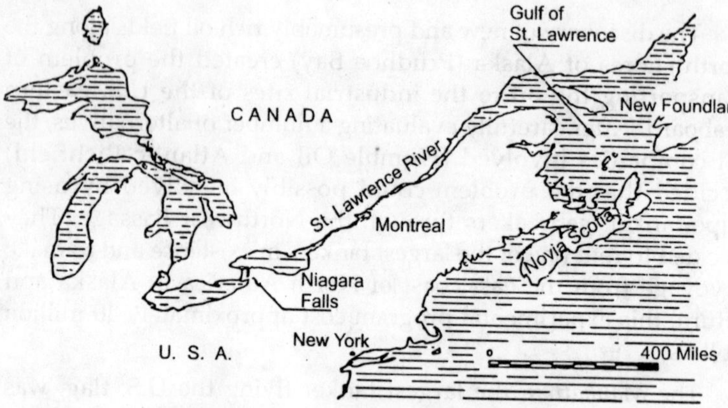

Fig. 11.1 : St. Lawrence seaway

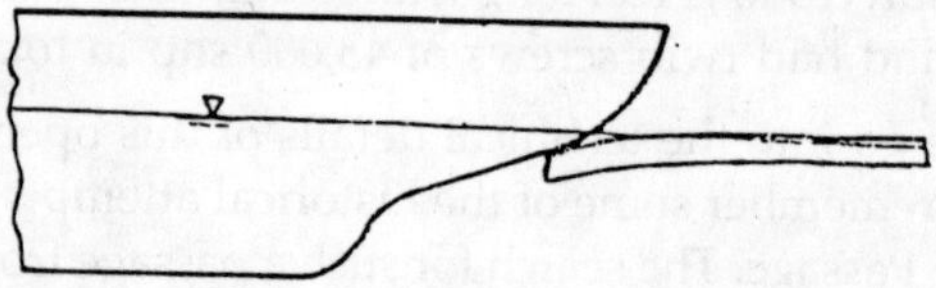

(a) Down-Breaking Type

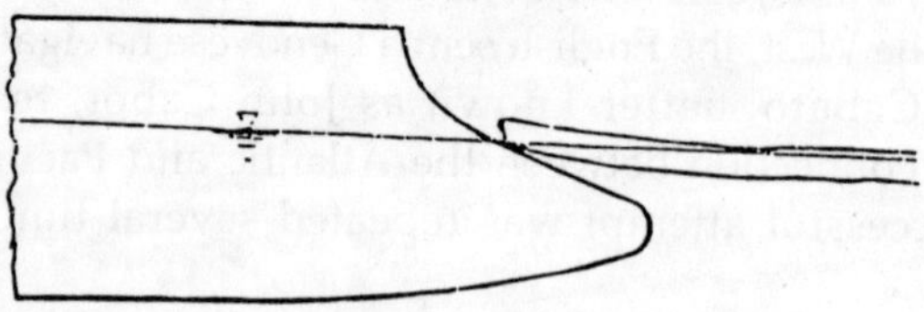

(b) Up-Breaking Type

Fig. 11.2 : Two types of icebreaking bows.

McClure rounded Banks Island from the west and demonstrated the possibility of a northwest passage, it was Roald Amundsen who made the first complete crossing from east to west, which took him three years. His voyage began in 1903.

The *Manhattan's* adventurous voyage into the icy arctic sea may mark the beginning of a new phase in the search for the Northwest Passage. The future will show the importance of this venture. Various aspects of the *Manhattan's* voyage are discussed in references 3 through 7.

Let us now look at the technical details that made the *Manhattan* suitable for this undertaking. It was clear from the beginning that the ship as it was would not be able to fight the arctic ice, and that important changes had to be made. The most essential change was her new icebreaking bow, which used the down-breaking principle. The bow inclines initially at 18° and bends to the conventional angle of 30° further down; in this way, icebreaking efficiency is increased by possibly 60%. In principle, two types of icebreaker bows can be used: the down-breaking bow and the up-breaking bow. Experience has shown that the down-breaking type is more effective. To reduce ice friction, the bow of the *Manhattan* was also widened to extend 8 feet beyond the main hull.

Other modifications included welding a 9-foot, 1 ¼-inch thick ice belt around the hull to strengthen it against ice forces, internal reinforcement with steel girders, strengthening of propeller shafts and screws, installing machinery collision shocks and constructing a helicopter flight check. The ship was also provided with a heeling system: the two bow thrusters serve as high-capacity ballast pumps, pumping 2000 tons of water from one side of the ship to the other in a maximum of 75 seconds, producing a rocking motion of 30° to both sides. It also included the installation of all types of modern equipment such as a satellite navigational system and computers that stored all data collected during the voyage.

The first severe test came as the Manhattan probed the heavy ice in McClure Strait, north of Banks Island. Winds sweeping across the open Arctic Ocean often jam the dense polar pack into the mouth of the strait, with heavy pressure and ridging, making it potentially the most difficult part of a transit. 'Faced with just this situation, the tanker was stuck in ice three times, and the decision taken to proceed via the more ice-free Prince of Wales Strait between Banks and Victoria Islands.

Proceeding via this alternate route she soon passed the Mackenzie delta and steamed up the Alaskan coast, arriving in sight of the drilling rigs at Prudhoe Bay on September 19. After onloading a symbolic barrel of Alaskan crude oil, she sailed on to Point Barrow, her successful westward transit completed.

On her return voyage the Manhattan spent almost a month in Melville Sound, deliberately seeking out different ice conditions. Maneuvering tests were run to see how large ships would perform under difficult circumstances comparable to those which would be encountered during much of the year, These tests concluded, she continued her eastward transit and arrived at New York City on November 12, having covered over 11,000 miles on the round trip.

Despite the reinforcements of the icebreaker, the *Manhattan* suffered a hull rupture due to ice impact, so that redesign of some hull characteristics will be required if additional experimental voyages are carried out. Because of the difficulties involved and the damage incurred, future explorations along this route are not at all certain. However, a vast amount of information was obtained on this voyage, which will be of great benefit to further studies.

The spectacular voyage of the *Manhattan* stresses thc great value of icebreakers in keeping coastal sea lanes open. We should realize that using icebreakers in the conventional way is still a rather crude approach and that definements will be necessary to further improve the system's efficiency such as possibly the additional use of explosives or the development of special tools such as ice-saws. Other methods of preventing navigation problems may usually be applied on a smaller scale. Among these are the use of air bubbling systems, special methods of cutting and removing ice, etc.

As an example, harbor basins in the port of Stockholm are kept open with an air bubbling system on the harbor bottom along the sides of the piers. It should be realized, however, that air bubbling systems are effective only when the flow generated by the system is water of a somewhat higher temperature from the basin depth. In salt water areas, the temperature gradient in vertical direction is usually very small and the air bubbling system consequently is less effective.

A large portion of northwestern Europe is favorably affected by the relatively warm water of the Gulf Stream, which runs along the east coast of the United States from Florida to Europe. It crosses the Atlantic Ocean at about 40° latitude and helps countries in northwest Europe keep their harbors and sea lanes open during the winter season. Even harbors at high latitudes benefit from it, such as the port of Hammerfest in northern Norway, which is open all year. Most Norwegian fiords which form many natural harbors are ice free.

Opposite the Scandinavian peninsula, the northern part of the Baltic is under less favorable conditions. This sea has a low salinity and is frozen during most of the year. The southern part of the Baltic Sea is in open connection with the North Sea through the Skagerak. This part is usually navigable, except in severe winters when ice conditions can seriously hamper navigation.

A ferry boat ride from the northern part of Germany to the southern tip of Sweden during a severe winter is a rewarding experience. During this trip the big ferry boat plows through the heavy ice most of its way.

The economic significance of thc problem of keeping sea lanes open has created a great interest in methods and techniques for solving this problem. The ice breaking techniques used in the United States and Canada have also been widely used by western Europe countries and Russia. Ice engineering plays a major role in Russia, as may be easily understood. The Russians are studying ice problems in both the laboratory and in the field and have developed powerful icebreakers.

In the Netherlands, ice breaking on rivers began in 1861 (8). The icebreakers were primarily used to break up ice fields and ice jams in securing the discharge of river flow toward the sea after severe winter conditions. The first ice breakers used on the Dutch rivers were steamboats; paddle wheel boats did a better job than boats with screw propulsion, Further experiments with icebreakers were carried out in the winters of 1875, 1879, 1880 and 1890. In the winter of 1890-91 the scale of operations was increased; of nine steamboats put into action, three were paddle wheel boats and six had screw propulsion. In the official report, however, the judgment on performance was still rather negative and the resultant effects of this operation seemed to have been rather disappointing.

In the severe winter of 1928-29 more powerful boats were put into operation and the results were indeed more successful. A similar experience during the winter of 1939-40 confirmed that for adequate ice breaking, under certain conditions, a minimum amount of horsepower is required. A relationship exists between ice thickness, the speed of the icebreaker crushing the ice and the amount of horsepower needed. The dotted lines are extrapolations based on observed data.

Only after 1940 were special requirements for ice breaking performance added to the ship. They consisted mainly of a uniquely designed bow, a stronger hull and adapted machinery and propulsion.

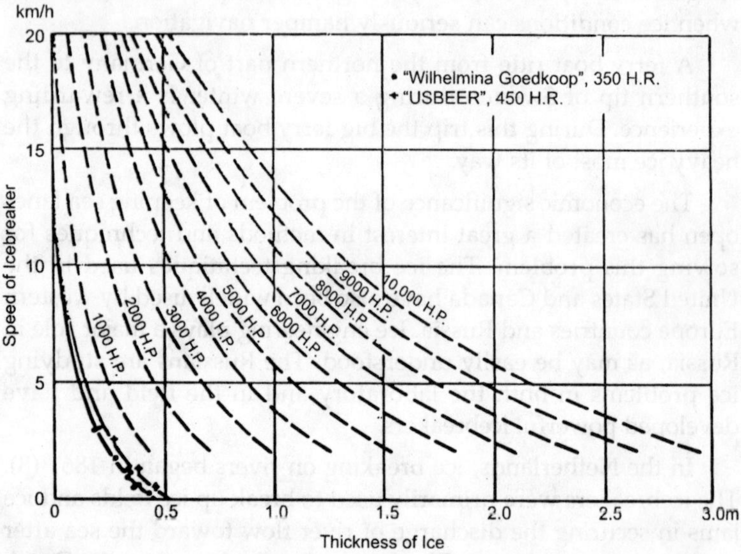

Fig. 11.3 : Relationship between speed of crushing ice, ice thickness and horsepower.

The breaking of ice for maintaining navigable conditions was essential for the ports of Amsterdam and Rotterdam. In 1929 several ice committees were organized to consider the possibilities of keeping navigation going as long as possible.

The water in Rotterdam Harbor is brackish and has a varying salinity with the tide phase. Therefore, chances for the formation of a solid ice deck are very slim, but the possibility of ice floes freezing together into large and heavy units does occur during severe winter conditions, and at that time the use of icebreakers has proven useful and economical. In the winter of 1962-63 a total of 1500 ice break hours were utilized to aid navigation in Rotterdam Harbor. The more powerful boats in the inland connection with Germany and also the powerful and adequately shaped push bows have been a great help in stimulating navigation traffic on the inland shipping routes.

11.2 The Delta Project

The Delta project is usually carried to protect the low-lying parts of southwest Holland against high storm tides.

It encompasses the closure of a number of large estuaries by building a dam from one bank to the other. In the Haringvliet Dam a large sluice is being built to enable the discharge of part of the Rhine runoff in cases of high river flow. An important consequence of the project has been the formation of a large fresh water lake south of the secondary dam through the Volkerak.

It is evident that such a major change in hydraulic conditions is likely to affect the discharge conditions of ice. In the past the Volkerak channel was a major route for the discharge of ice toward the sea due to prevailing easterly and northeasterly winds, but it is now no longer possible to use this discharge route. Consequently, ice has to be discharged through the 17 openings of the Haringvliet Sluice, about 1 kilometer long or through the estuary sections near Rotterdam. Water circulation studies on the Haringvliet have made clear that the wide Haringvliet Estuary will not be the best way to discharge ice because of high ice productions on the Haringvliet itself, which has fresh water and long periods of low velocities.

Measures to improve discharge of ice through the Haringvliet could be the following:

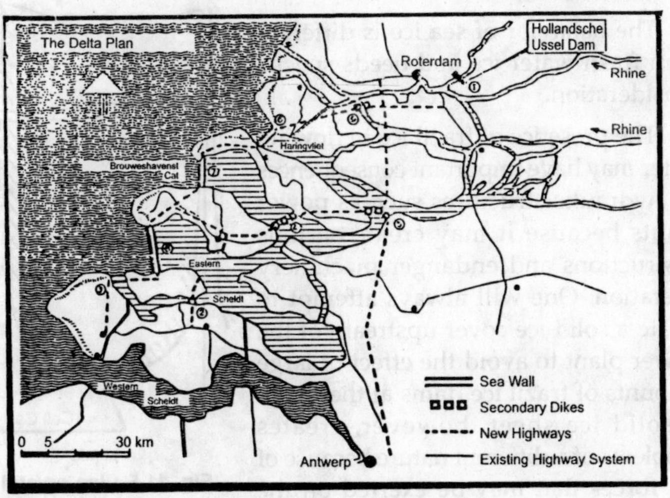

Fig. 11.4 : Delta Project

- Opening the sluices during ice conditions and reestablishing a tidal motion (sea water) on the Haringvliet;
- Opening the sluices during one or more ebb periods, thereby increasing the drift velocity in seaward direction.

The first possibility did not create sufficient relief to justify the emporary reestablishment of a salt water estuary in the Haringvliet (with all its disadvantages). The second solution was not accepted because manipulating the sluices in the suggested way would intolerably increase salt water conditions on the Rotterdam waterway.

Calculations have been carried out by means of an electric analog computer (representing the entire estuary system) to determine whether the discharge of ice along the northern branches is a hydraulic feasibility or whether other ways should be found to cope with the current problems. The computations have shown that ice discharge along the New Waterway seems to be a feasible solution, although less desirable from the standpoint of navigation. Discharge along the routes of Lower Merwede, Oude Mass and New Waterway is less promising because of small resultant ebb drift velocities in the neighborhood of Dordrecht. Further studies and full-scale experimentation will be required to arrive at a satisfactory solution of the problem.

11.3 The Behavior of Sea Ice

The behavior of sea ice is different from fresh water ice and needs special consideration.

The presence of frazil ice in flowing water may have important consequences for hydraulic structures such as power plants because it may create serious obstructions and endanger machinery operation. One will always attempt to create a solid ice cover upstream of the power plant to avoid the effect of large amounts of frazil ice dams at the plant. A solid ice sheet, however, creates problems of a different nature because of the forces that may be exerted on the structure.

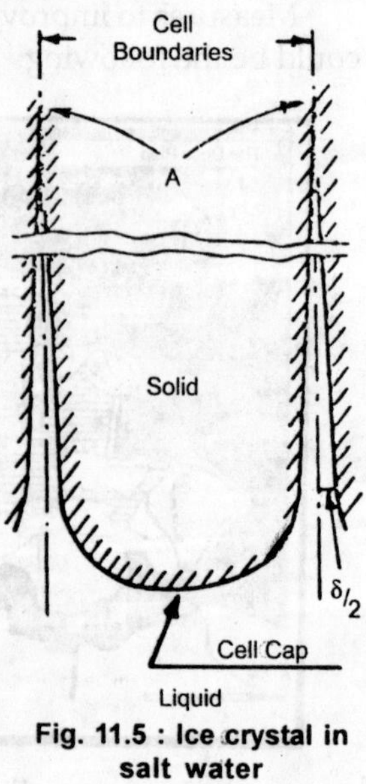

Fig. 11.5 : Ice crystal in salt water

Ice is crystalline material. When loads are applied to it, the material shows both elastic and plastic behavior. The plastic behavior is directly related to the crystal structure. It is usually built up of long, pencil-like grains and is then called columnar ice. In snow ice the crystals have more of a granular structure and this is called granular ice. Crystalline materials like ice deform plastically by slip on preferred crystallographic planes. The plane perpendicular to the axis of hexagonal symmetry of the ice crystal (the so-called C axis) is called the basal plane; it plays a very important role in the plastic deformation (2). Slip in ice occurs most easily on the basal plane. In other planes much higher shear stresses are required to cause deformation, depending on the value of the shear stress.

Because of the combination of plastic and elastic properties, the stresses induced in ice masses are dependent on time-load relationships. This has a considerable influence on the test results that have been applied and accounts for a great deal of wide scatter in values given by different authors for the elastic constants of ice. Furthermore, both the elastic and plastic properties of ice are time dependent. Because the formation of ice is also dependent on weather conditions, we may expect to trace some ice properties back to changes in these conditions.

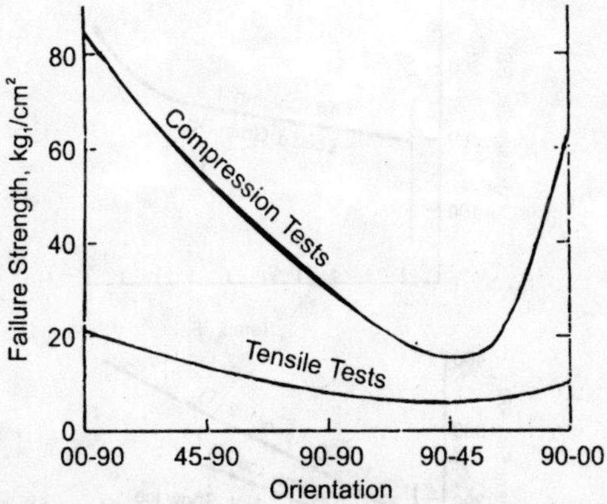

Fig. 11.6 : Dependency of strength on crystal orientation (after Peyton, 1966).

Elastic properties are determined by the value of the elastic modulus E. As observed before, the value of E will be dependent on the loading characteristics. The elastic properties prevail when no time is given to plastic deformation. For periods of loading less than about 10 seconds, and depending on the temperature, the deformation of ice is almost entirely recoverable and can be considered elastic. For periods greater than 10 seconds the ice will undergo plastic deformation (3). The value of the "dynamic" E can be obtained by studying dynamic effects of small deflection such as the propagation of a shock wave through an ice cover.

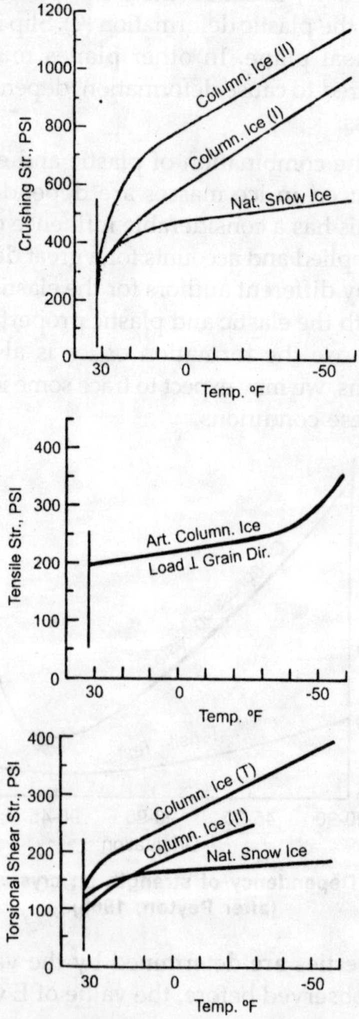

Fig. 11.7 : Elastic properties of ice as a function of temperature (after Butkovich, 1954)

11.4 The Forces Exerted by Ice

The forces exerted by ice on structures depend on a great many variables. The important variables can be explained as follows:

- Movement of ice cover relative to the structure;
- Flexibility of the structure;
- Characteristics of the ice cover and
- Properties of the ice.

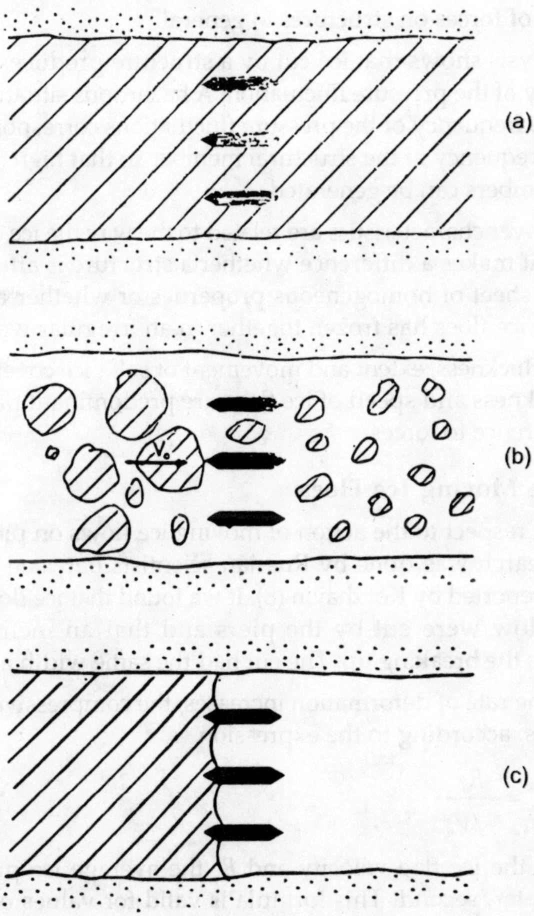

Fig. 11.8 : Different load situation for bridge.

The first factor plays an essential role in large ice fields on lakes and reservoirs, where movements of large areas of ice may take place, often due to unpredictable circumstances. When a structure is in the middle of such an unbroken ice area, ice field movements may cause great forces on it. Maximum forces will probably be found by multiplying the maximum crushing pressure of ice by the area of contact between the ice sheet and the structural elements upon which the ice force is working. The total forces are also very much influenced by the flexibility characteristics of the structure on which the ice force is working. This is similar to the problem of forces on structures in general.

Analysis shows that ice cut by a structure produces a certain frequency of the pressure fluctuation. A hazardous situation occurs when the frequency of the pressure fluctuations corresponds to the natural frequency of the structural member so that high stresses in these members can be generated.

Ice cover characteristics are related to the way the ice cover was formed; it makes a difference whether a structure is affected by a solid ice sheet of homogeneous properties or whether a cover of irregular ice floes has frozen together in an irregular way.

The thickness, extent and movement of solid ice covers and the size, thickness and speed of ice floes are predominant parameters with reference to forces.

11.5 The Moving Ice Floes

With respect to the action of moving ice flows on piers, much field research was done by Russian scientists between 1934 and 1958, as reported by Korzhavin (8). It wa found that ice floes carried by the flow were cut by the piers and that an inclined nose facilitates the breaking up. The cut had the same width as the pier.

As the rate of deformation increases, the compressive strength decreases, according to the expression

$$R_{compr.} = \frac{R_1}{\sqrt[3]{V_0}}$$

V_0 being the ice floe velocity and R_1 the average ice pressure at $V_0 = 1$ meter/second. This formula is valid for values of $V_0 > 0.1$ meter/second. For R_1, a value of 50 ton force/meter is indicated, which seems rather low compared to an average compressive strength of ice of 280 ton force/meter^2.

However, local compression due to pier penetration in the ice floe was found to increase interaction forces two to two and a half times. A form factor (m) for a pier with a triangular form was suggested, as given by the expression m = 0.39 (1 +α), α being one-half of the pier nose angle.

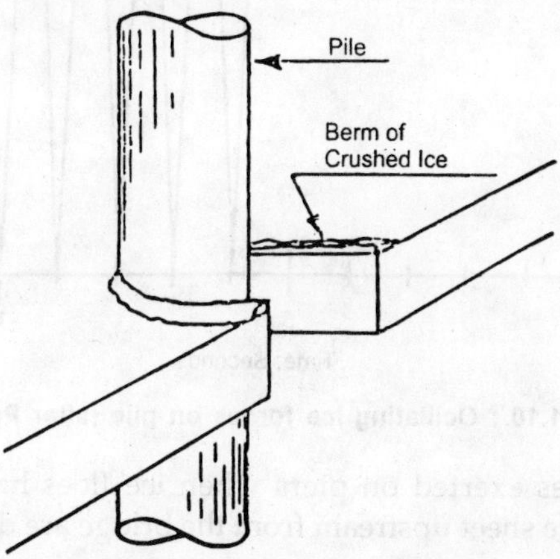

Fig. 11.9 : Cutting of ice by pile (after Peyton).

In Canada, design engineers have been using the average crushing strength (280 ton force/meter^2) as a basis for their design of highway bridges across rivers. Field measurements, obtained by pressure devices into two bridge piers over a four-year period have yielded unit pressures from moving ice-sheets in the range of 70-113 ton force/meter^2 (7 to 11.3 kgf/cm^2) for ordinary to moderate conditions. The value of 280 ton force/meter^2 specified in Canadian codes may therefore be too conservative.

This criterion presumably gives a safe upper limit to the load, but the amount of safety is unknown and probably unnecessarily large compared with the Russian figures.

Some Canadian designers have accepted the values used in dam design, ton force/meter. With an average thickness of 2 feet (0.6 meter), this corresponds to 50 ton force/meter^2 , which comes close to the figures mentioned in the Russian publication and also corresponds to figures based on thermal expansion.

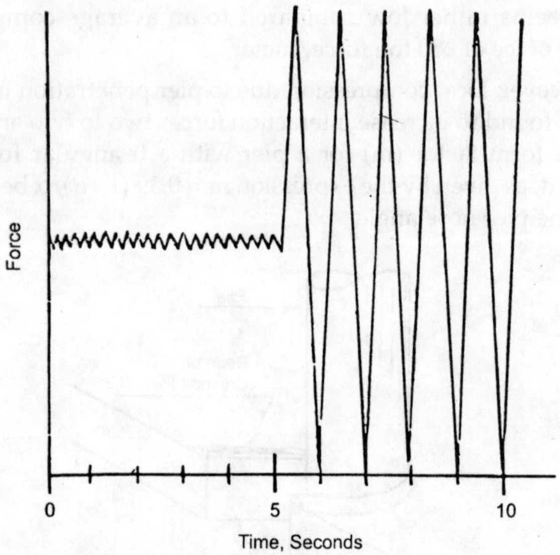

Fig. 11.10 : Ocillating ice forces on pile (after Peyton)

The forces exerted on piers when ice floes have formed a continuous ice sheet upstream from the bridge are depend on;

- The size of the continuous ice sheet;
- Roughness of underside ice deck and
- Hydraulic conditions on the river (bottom roughness, depth, slope).

If the ice deck connects the banks, the latter may exert shearing forces on the ice deck and reduce the forces on the piers.

Forces on the piers become maximum when the ice is loose from the banks. They can be computed based on the friction that the water flow underneath the ice deck exerts on the ice deck.

The presence of bridge piers in a river may give rise to the formation of a local ice jam. In that case a head difference may be formed, exerting an additional horizontal force on the hanging ice dam and consequently on the piers.

In a given situation the conditions should be analyzed to determine whether or not such an ice dam can be formed or whether its formation can be corrected by carrying out preventive measures.

Chapter Twelve
The Current Systems in the Oceans

12.1 The Indian Ocean Basin

The Indian Ocean basin is the smallest of the five great subdivisions of the world ocean with $49.10^6 km^2$ out of the $361.10^6 km^2$ of the global ocean. It is closed to the north around the latitude of the Tropic of Cancer by the Asian continent, which has important consequences on the ocean circulation. South of the equator, its western boundary is modified by the presence of the island of Madagascar. In the east, the basin is connected with the equatorial Pacific Ocean through the deep passages of the Indonesian Seas. The north of the Indian Ocean is made up of the large basins on either side of the Indian peninsula, the Arabian Sea in the west and the Bay of Bengal in the east which drains most of the river runoff from the Himalayas and the Indian subcontinent. The Arabian Sea is connected directly to the shallow Persian Gulf, and through the sill of Bab-el-Mandeb (110m) to the deep Red Sea basin where high salinity waters are formed. In the south, the basin is largely open to the Antarctic Ocean between South Africa and Australia. The Indian Ocean limit to the south is the Subtropical Convergence, a hydrological limit where the meridional surface temperature gradient is maximum. At depth, the complicated system of ridges separates the Indian Ocean in many deep basins.

Due to the presence of the Asiatic continent to the north, the atmospheric circulation is quite different from the Pacific Ocean and the Atlantic Ocean, particularly north of 10°S. Seasonal heating and cooling of the atmosphere over Asia induces a seasonally varying monsoon circulation (Figure 2). For centuries it has been known that the winds north of around 10°S reverse with the seasons. A long time ago the Arabic traders along the east African coast made use of the fair currents and winds during their voyages. The word 'monsoon' comes from the Arabic word 'mawsin' meaning season. As the winds are the main driver of the currents, in particular near the surface, the main characteristics of the wind seasonal variability will be described below.

The wind seasonal variability over the ocean can be separated in four periods: the winter monsoon period, the summer monsoon period, and the two transition periods between the two monsoons.

Between December and March–April, north of the equator, the winter (NE) monsoon blows from he north east with a moderate strength. At the equator the winds are weak and usually from the north. Between the equator and the Intertropical Convergence Zone (ITCZ) which stretches zonally near 10ºS between north Madagascar and south Sumatra, it blows from the north west. During that season (southern summer), the atmospheric pressure decreases over Australia and South Africa and the subtropical high pressure over the ocean, around 35ºS, is weaker – as are the south-east trade winds during that season.

In April–May, during the transition period between the end of the NE monsoon and the beginning of the SW monsoon, the winds north of the equator calm down. At the equator moderate east-ward winds blow, which contrasts with the west-ward winds over the equatorial Pacific and Atlantic Oceans.

From June to September–October during the SW monsoon, the winds reverse completely and north of the equator the summer monsoon blows steadily from the south west. The SW summer monsoon is much stronger than the NE winter monsoon. A wind jet, also called the Finlater jet, develops along the high orography of the east African coast. As a consequence, the winds are the strongest on the western side of the Indian Ocean along the Somali coast towards the Arabian Sea, particularly north-east of Cape Guardafui (the horn$_{-1}$ of Africa) where the mean July wind speed is $12 ms^{-1}$ with peaks exceeding $20 ms^{-1}$. They are the strongest and the steadiest wind flow in the world. At the equator, the winds are moderate from the south and decrease eastward. In the southern Indian Ocean, the subtropical high pressure center intensifies and covers the whole width of the southern Indian Ocean during the southern winter (July) and the SE trades penetrate farther north than during the southern summer (January); they reach the equator in the western part of the ocean and are the strongest among the three oceans. During that season the air masses transported by the SE trade winds cross the equator in the west and continue, loaded with moisture, towards the Asian continent where they bring the awaited monsoon rainfall.

October–November corresponds to the second transition period between the end of the SW monsoon and the beginning of the NE monsoon. North of the Equator, the winds vanish and the sea surface temperature can exceed 30°C. At the equator moderate eastward winds blow again as during the first transition period, although they are usually slightly stronger. The Indian Ocean is the smallest of all the oceans and is in several respects quite different from the others. In particular, it is bounded by the Asian continent to the north. This meridional land–sea contrast has a strong influence on the winds, resulting in a complete seasonal reversal of the winds known as the monsoon system. The characteristics of the basin and of the wind regime are determinant for the currents, and will be described first in this article. The description of the currents has been separated into two main sections: the first for the southern part of the Indian Ocean which is not affected by the monsoons and is more akin to the other subtropical oceans; and the second for the northern part which undergoes forcing through the reversal of the monsoon winds.

As in the other oceans, the strongest currents are close to the western shores of the ocean as a result of the direction in which the earth rotates and the variation of the Coriolis parameter with latitude. North of 10°S (which is the limit of the monsoon influence), there are two western boundary currents: the East African Coastal Current (EACC) which always flows northward and the Somali Current which is the most intense and the most variable.

The EACC flows in continuity with the branch of the SEC which passes north of Madagascar and splits around 11°S. It runs northward throughout the year between latitudes 11°S and 3°S. Its surface speed can exceed $1 ms^{-1}$ during northern summer and its transport amounts to 20Sv in the upper 500m. Its northern end depends on the season. In northern winter, the EACC converges around 3°S–4°S with the south-going Somali Current to form the eastward South Equatorial Countercurrent. During the northern summer, the EACC merges into the north-going Somali Current.

The Somali Current is the most intense, but unlike the other western boundary currents it is highly variable due to the complete seasonal reversal of the winds. It was first studied during the International Indian Ocean Experiment in the 1960s, during INDEX (Indian ocean experiment which started in the 1970s) and during SINODE in the 1980s. Recently (1990–96), during WOCE (World Ocean Circulation Experiment) considerable amounts of new data were collected over the whole Indian Ocean. The Somali current develops in different phases in response to the winds.

During the transition period after the NE monsoon, in April, the Somali current flows south-westward along the coast south of 5° N, merging near the equator with the northward-flowing EACC. This feeds a south-eastward flow towards the ocean interior.

In early May, the Somali current responds rapidly to the onset of the local southerly winds and reverses northward in continuity with the northward EACC near the equator. By mid-May, the SW wind onset propagates northward and the current turns offshore towards the east at 4°N. North of that branch an upwelling wedge spreads out, bringing cold and enriched waters at the surface. Further to the north, the current flows northward from March onwards. When the onset of the strong summer monsoon winds occurs at these latitudes in June, the southern branch increases in strength and a strong anticyclonic gyre, called 'the great whirl' develops between 5°N and 10°N. Between the Somali coast and the northern branch of the great whirl a second upwelling wedge forms. Numerical models have shown that the location and motion of these structures are influenced by the distribution and strength of the wind forcing.

In August–September, when the winds decrease, the southern cold wedge propagates northward along the coast and meets with the northern one (although the latter probably also moves). It is only at that time that the Somali current is continuous from the equator up to 10°N and brings fresher waters into the Arabian Sea.

During the transition period in October–November, the northward Somali circulation decreases.

In December–February, during the NE winter monsoon, the Somaii Current reverses southward from 10°N to 5°S where it converges with the northward EACC to form the South Equatorial Counter Current (SECC) flowing eastward. This countercurrent exists only during the NE winter monsoon and could be compared to the other Equatorial Counter Currents in the Atlantic Ocean and in the Pacific Ocean. It develops just north of the intertropical convergence zone (ITCZ) where the winds have an eastward component.

At the equator near Africa, the reversal affects only a thin surface layer below which (between 120m and 400m) there is a northward undercurrent, remnant of the SW monsoon season, followed again by a southward current below 400 m.

There are also western boundary currents in the Gulf of Bengal off the coasts of Sri Lanka and India.

From September to January, the currents are southward along the whole eastern coast of India and Sri Lanka, bringing fresh Bengal Bay water to low latitudes (6°N). In February–March the currents reverse to flow to the north along these coasts with a separation from the coast in the northern Bay of Bengal (19°N) in March. From April to August, the current reverses along the eastern coast of Sri Lanka where it flows to the south. Further north, from May to July, the separation from the coast of the northward current takes place around 16°N instead of 19°N. In July, north of 16°N, reversal to the south takes place.

12.2 The Atlantic Ocean Currents

The voyages of discovery brought interesting observations of many of the important surface currents of the Atlantic Oceans.

Recently, much attention has been focused on climate research as it became clear that climate goes through long period variability and can affect our lives and prosperity. The physical climate system is controlled by the interaction of atmosphere, ocean, land and sea ice, and land surfaces. To understand the influence of the ocean on climate, the physical processes and especially the ocean currents storing and transporting heat need to be thoroughly investigated. Since the end of the twentieth century, the general circulation of the Atlantic Ocean has been considered in a climatological context. A new picture emerged with me North Atlantic Ocean being seen not only as relevant to the climate of Europe, but for its influence on the entire globe due to its unique thermohaline circulation. Its warm upper-ocean currents transport mass and heat, originating in part from the Pacific and Indian Oceans, towards the north in the South-Atlantic and the North Atlantic. Cold deep waters of the North Atlantic flow southward, cross the South Atlantic, and are exported into the Indian and even the Pacific Oceans. The research focus on ocean currents in me Atlantic Ocean at the beginning of the new millennium will be further investigation of the mean surface and deep currents of the Atlantic and its variability on short-term to decadal timescales, so that the ocean's role in climate change can be better understood.

The Mid-Atlantic Ridge, which in many parts rises to <2000m depth and reaches the 3000m depth contour nearly everywhere, is located zonally in most places near the middle of the Atlantic and divides the Atlantic Ocean into a series of eastern and estern basins. The basin names presented in Figure 1 are only the major ones; in more detailed investigations of special regions many more topographically identified structures exist with their related names. The major topographic features of the Atlantic strongly affect the deep currents of the deep and bottom water masses, either by blocking or guiding the flow. The Walvis Ridge off south-west Africa limits and northeward flow of Antarctic Bottom Water (AABW) in the Cape Basin and consequently the major northward flow of AABW takes place in the western basins. Although the Rio Grande Rise between the Argentina Abyssal Plain and the Brazil Basin disturbs a smoothe northward spreading of the AABW in the western basins, the Vema and Hunter Channels within the Rio Grande Rise are deep and wide enough to allow a continuous northward flow. Te Romanche Fracture Zone at the equator allows part of the AABW to enter the eastern basins of the Atlantic, where the AABW spreads poleward in both hemispheres.

Charts of ocean currents from the late nineteenth century show that by then the patterns of surface circulation in regions away from the equator and polar latitudes were already well understood. This fundamental knowledge accumulated gradually through centuries of sea travel and had reached a state of near correctness by the time dedicated research cruises, full depth measurements, and the practical application of the dynamical method were begun.

By the fifth century AD, mariners had probably acquired intimate knowledge of coastal currents in the Mediterranean, but little information abour them is reported in Classical writing. Following the dark and Middle Ages when little progress was made, the voyages of discovery brought startling observations of many of the Atlantic's most important ocean currents, such as the North and South Equatorial Currents, the Gulf Stream, the Agulhas Current, and others. The Gulf Stream appears to have been mapped as early as 1525 (by Ribeiro on the basis of Spanish pilot charts. The fifteenth to seventeenth centuries were marked by attainments of knowledge that increasingly taxed the abilities of science writers to reconcile new information with accepted doctrine.

The warmwatersphere, consisting of the warmer uppet waters of the ocean, is the most climatologically important part of the ocean due to its direct interaction with the atmosphere. The transition from the warm- to the cold-watersphere takes place in a relatively thin layer at temperarures between 8° and 10°C. The warmwatersphere reaches to 500–1000m depth in the Atlantic's subtropics and rapidly rises towards the ocean surface poleward of about 40° latitude.

The near-surface circulation is driven primarily by the wind and forced into closed circulation cells by the continental boundaries. The circulation of the Atlantic is governed by the subtropical gyres of the North and South Atlantic. The subtropical gyre of the North Atlantic includes the Florida Current and Gulf Stream as western boundary currents, the North Atlantic Current, the Azores and Canary Currents in the easrern Atlantic, the North Equatorial Current, and the Caribbean, Cayman and Loop Current in the Caribbean and Gulf of Mexico. The subtropical gyre of the South Atlantic includes the poleward-directed Brazil Current as western boundary current, which turns eastward at the Brazil Falkland (Malvinas) confluence region as eastward flow near 40°S named South Atlantic Current. The South Atlantic Current in part continues to the Indian Ocean and in part adds to water from the Agulhas retroflection to the northward-flowing Benguela Current and the westward-flowing South equatorial current.

Near the coast of north Brazil the South Equatorial Current contributes in part to the Brazil Current, but in part also to the subsurface intensified North Brazil Undercurrent, responsible for the warm water flow from the Southern to the Northern Hemisphere. The two anticyclonic subtropical gyres, clockwise in the Northern and counterclockwise in the Southern Hemisphere, reach through the entire warmwatersphere and show only weak seasonal changes. Subtropical gyres, although existing longitudinally to basin-scale, also tend to have sub-baisn-scale recirculation gyres in their western reaches. The northward extent of the South Atlantic subtropical gyre decreases with increasing depth. It is located near Brazil at 16°S in the near-surface layer and at 26°S in the layer of Antarctic Intermediate Water.

The preferential north–south orientation of the continents bounding the Atlantic Ocean lead to meridional eastern and western boundary currents which together with the wind-induced zonal currents, westward flow under the trade winds, and eastward flow under the midlatitude westerly winds, form the closed gyres. The western ocean boundary regions are associated with an intensification of the currents. The consequent energetic western boundary currents, the Florida Current and the Gulf Stream in the North Atlantic Ocean and the Brazil Current in the South Atlantic Ocean, have large transports and typical width scales of ~100km. The western boundary currents are intensified because the strength of the Coriolis effect varies with latitude. The western boundary currents of the Atlantic Ocean are generally so deep that they are constrained against the continental shelf edge and do not reach the shore (*see* **Brazil and Falklands (Malvinas) Currents and Florida Current, Gulf Stream and Labrador Current**).

In the Tropics the two subtropical gyres are connected via a complicated tropical circulation system. The tropical circulation shows a north-westward cross-equatorial flow at the western boundary and several zonal current and countercurrent bands of smaller meridional and vertical extent and a lot of vertical diversification. The north-westward flow along the western boundary starts as a subsurface flow, the North Brazil Undercurrent, which becomes surface intensified north of the north-eastern tip of Brazil by near-surface inflow from the South Equatorial Current and is named North Brazil Current. The North Brazil Current crosses the equator north-westwards and retroflects eastward at about 8°N. In this North Brazil Current retroflection zone, eddies detach from the current and progress north-westward towards the Caribean.

In northern spring, when the North Equatorial Countercurrent is weak, there seems to be a continuous flow of about 10 Sv towards the Caribbean called Guyana Current. The westward flows are regarded as different bands of the South Equatorial Current, the northern one even crossing the equator. The eastward subsurface flows are named the Equatorial Undercurrent at the equator, and the North and South Equatorial Undercurrents at about 5° latitude. The eastward surface intensified flows at about 9° latitude are the North and South Equatorial Countercurrents. In northern fall the North Equatorial Countercurrent and the North Equatorial Undercurrent override one another and it is difficult to distinguish between the two current bands. In the Antarctic Intermediate Water layer at about 700 m depth there are intermediate currents at the equator (Equatorial Intermediate Current), as well as north and south of the equator (Northern and Southern Intermediate Countercurrents) which flow in the opposite direction to the currents above. The Intertropical Convergence Zone in the Atlantic, where the trade winds of both hemispheres converge, is located north of the equator throughout the year, and reaches the South Atlantic only in southern summer and then only at the north coast of Brazil. Seasonal changes of the wind field lead obvious variations in the tropical near-surface currents; however, with different strengths. The strongest seasonal signal is observed in the North Equatorial Countercurrent. The eastward-flowing North Equatorial Countercurrent is strongest in August, when the Intertropical Convergence Zone is located at its northernmost position. At that time the North Equatorial Countercurrent crosses the entire Atlantic basins zonally, but in late boreal winter it becomes weak or even reverses to westward in the western domain. South of the Cape Verde Islands at 9°N, 25°W, there is a cyclonic feature named Guinea Dome throughout the year, but it is weaker in northern. winter. The Southern Hemispheric counterpart is the Angola Dome at 10°S 9°E. The Angola Dome is seen only in southern summer and it is imbedded in a permanent larger-scale cyclonic feature centered near 13°S, 5°E called Angola Gyre.

The western tropical Atlantic is a region of special interest in the global ocean circulation. The meridional heat transport across the equator is accomplished by warm surface water, central water, and subpolar intermediate water from the Southern Hemisphere moving northward in the upper 900 m mainly in the North Brazil Current, and cold North Atlantic Deep Water (NADW) moving southward between 1200m and 4000m. These reversed and compensating water spreading paths are often referred to as part of the global thermohaline conveyor belt. A clear distinction has to be made between the cross-equatorial flow at the western boundary and the interhemispheric water mass exchange. The latter is the amoun of transfer from the Southern to the Northern Hemisphere of about 17 Sv in the upper ocean, and to a small degree in the Antarctic Bottom Water, compensated by the transfer from the Northern to the Southern Hemisphere of about 17 Sv by the North

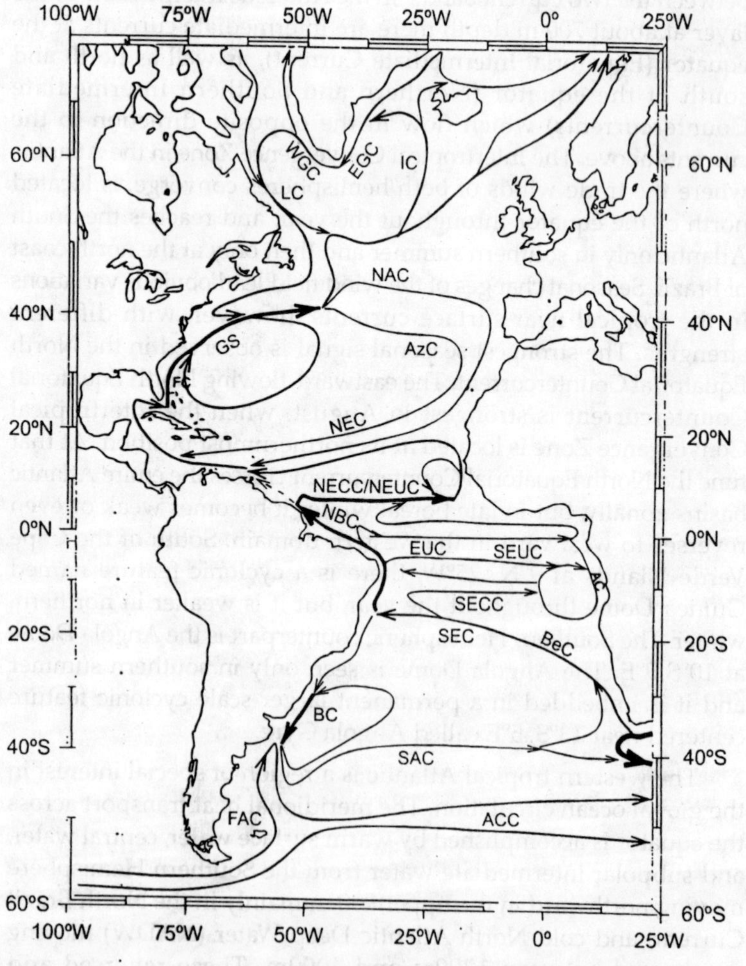

Fig. 12. 1 : Schematic representation of upper-ocean currents in the North and South Atlantic Oceans

Atlantic Deep Water rent with about 35 Sv is much larger, since part of this cross-equatorial flow originates from the zonal equatorial circulation, retroflects north of the equator, and returns into the equatorial circulation system.

Poleward of the Subtropical gyre the current field of the North and South Atlantic are completely different. In the North Atlantic a cyclonic subpolar gyre is present, driven in part by the wind stress curl associated with the atmospheric Icelandic low pressure system and in part by the fresh water from the subarctic. This subpolar gyre includes the northern part of the North Atlantic Current, the Irminger Current, the East and West Greenland Currents and the Labrador Current off north-eastern North America. In the South Atlantic the continents terminate and an eastward flow of water all around the globe within the Antarctic Circumpolar Current is driven mainly by the midlatitudes westerlies. The South Atlantic Counterpart of the Labrador Current is the Falkland (Malvinas) Current, which flows equatorward along the south-eastern South American shelf edge to about 38°S. However, this current differs in origin as it is essentially a meander of a branch of the Antarctic Circumpolar Current. In the Brazil/Falkland (Malvinas) confluence region the Falkland (Malvinas) Current is retroflected southward to join the Antarctic Circumpolar Current. The South Atlantic Current as southern current band of the South Atlantic subtropical gyre and the Antarctic Circumpolar Current can be distinguished as separate current bands, nevertheless mass and heat exchange between the subtropics and subpolar region takes place in this region.

Now let us study in brief about the currents of deep Atlantic Ocean. The deep-ocean curculation depeneds heavily on the changes in density imposed by air-sea interaction. The flow in the deep ocean is driven by the equator-to-pole differences in ocean density. This thermohaline circulation, driven by temperature and salinity gradients, provides global-scale transport of heat and salt. The forcing of this flow is concentrated in a few areas of intense production of dense water in the far North Atlantic, the Labrador Sea, and along the margin of Antarctica within so-called convection areas.

To improve the climate prediction models it is necessary to understand the ocean's role in climate changes better. International programs like Climate Variability and Predictability (CLIVAR) started to describe and understand the physical processes responsible for climate and predictability on seasonal, interannual, decadal, and centennial timescales, through the collection and analysis of observations and the development and application of models of the coupled climate system.

Another new important focus will be to describe and understand the interactive physical, chemical, and biological processes that regulate the total Earth system. This is also the overall objective of the International Geosphere-Biosphere Program (IGBP). One core project of IGBP is GLOBEC, which is now changing from a planning to a research status with the goal of advancing understanding of the structure and functioning of the global ocean ecosystem, its major subsystems, and its response to physical forcing so that a capability can be developed to forecast the responses of the marine ecosystem to global change.

Despite the future focus on the Atlantic's role in climate changes as well as interactive processes, and although the major components of the near-surface circulation from ship drift observations have been known for more than 100 years, cltere is still also the need to investigate details of the Atlantic Ocean subsurface and abyssal circulation and its physical processes, which so far are unrevealed.

12.3 The Nature of Southern Ocean Currents

The southern ocean encircling Antartica plays a major role in shaping the characteristics of the global ocean.

It provides the most significant inter-ocean conduit by which waters of the three major oceans, the Atlantic, Pacific and Indian, can intermingle, acting to diminish their differences in temperature, salinity, and chemical properties. The most prominent current is the Antarctic Circumpolar Current, which transfers about 134 million m^3s^{-1} of sea water from west to east within a latitudinal range from 50° to 60°S. North of the Antarctic Circumpolar Current are the poleward limbs of the large subtropical gyres of the southern hemisphere, referred to as the South Atlantic, South Indian, and South Pacific Currents. Within large embayments of Antarctica, notably the Weddell and Ross Seas, south of the Antarctic Circumpolar Current are large clockwise flowing gyres. The Weddell Gyre carries about 30–50 million m^3s^{-1} of water. Along the continental margin of Antarctica is the coastal current that advects water from east to west. The coastal current is directed towards the north along the east coast of Antarctic Peninsula, forming the western boundary of the Weddell Gyre.

At the northern tip of the Antarctic Peninsula the coastal current is directed into the open ocean. The coastal waters injected into the open ocean separate the Antarctica Circumpolar Current from the Weddell Gyre, in what is called the Weddell-Scotia Confluence. Besides ocean currents flowing on nearly horizontal planes, the Southern Ocean experiences major overturning of ocean water. Overturning is forced by the production of dense surface water along the margins of Antarctica, leading to the formation of Antarctic Bottom Water.

Ocean currents are the product for the most part on the stress exerted on the sea surface by the wind. The winds are strong over the Southern Ocean, particular within the Indian Ocean and Australian sectors and therefore drive a vigorous circulation. Strong westerlies (wind directed from west to east) extend from the subtropical high atmospheric pressure near 30ºS, a latitude often used to define the northern limits of the southern Ocean, to a belt of low atmospheric pressure at 65ºS. South of 65ºS the winds are easterlies, marking the northern edges of the polar high pressure over Antarctica.

Density changes of surface water induced by sea-air fluxes of heat and fresh water, often involving sea ice within the Southern Ocean, also produce circulation. However, buoyancy (sometimes referred to as thermohaline), circulation is sluggish and mainly occurs within the meridional vertical plane, as dense water sinking forces slow, compensatory upwelling of less dense resident water. Sinking of dense waters along the continental margins of Antarctica results in Antarctic Bottom Water.

Ocean currents are for the most part in equilibrium wide the distribution of density within the ocean satisfying approximately the so-called 'thermal wind equation' (*see* **Elemental Distribution: Overview; Ocean Circulation**). Along the Greenwich meridian the surface of equal density, or isopycnals, rise up towards the sea surface as latitude increases to the south. Strongly sloped isopycnals are coupled to strong ocean current, more precisely to strong geostrophic ocean currents, relative to the seafloor. As a rule of

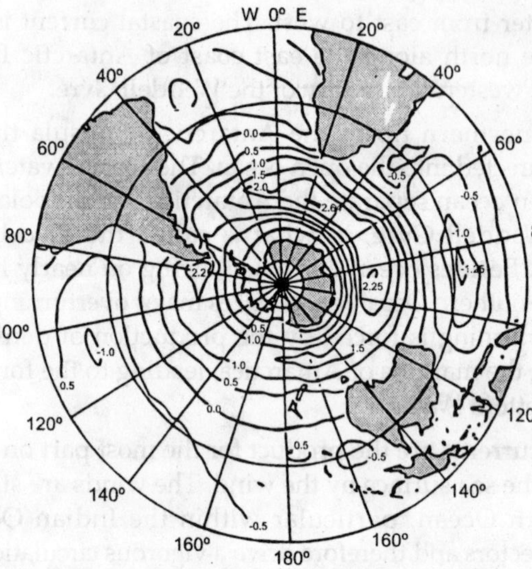

Fig. 12. 2 : Wind stress in Nm^{-2}. (Reproduced with permission from Nowlin and Klinck, 1986)

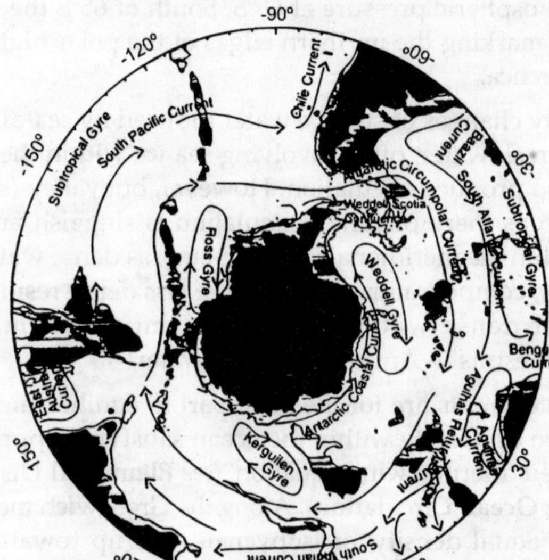

Fig. 12.3 : Schematic of general circulation of the Southern Ocean. Land is black; the shaded area marks ocean depth of <3000 m.

thumb, in the southern hemisphere higher density water occurs to the right of the direction of the ocean current. Hence the increasing density as latitude increases is linked to west to east flow of water. Along the margin of Antarctica the descent of isopycnals marks a flow towards the east. Regions of rapid changes in temperature, salinity or density mark the positions of ocean fronts, often coinciding with a strong ocean current.

Maximum westerlies in the wind occur near 55°S, which roughly coincides with the axis of the Antarctic Circumpolar Current. To the south of this latitude, wind-induced northward

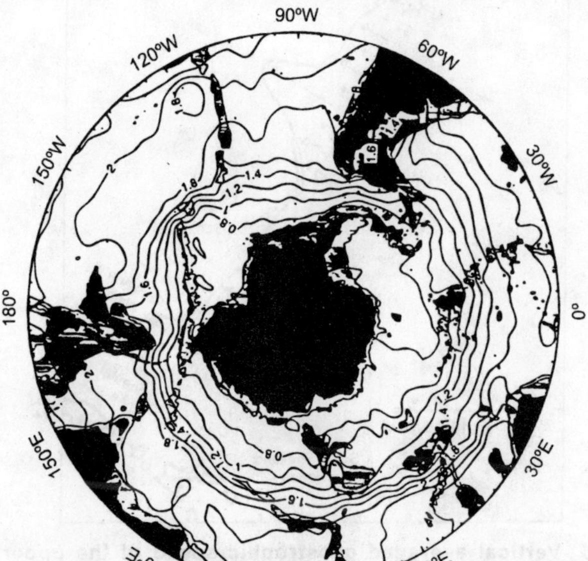

Fig. 12.4 : Anomaly of Southern Ocean sea level height relative to the 2×10^7 Pa pressure surface. The value are differences in dynamic meters (approximately equivalent to geometric meters) from that of a standard ocean (0°C, 35 PSU salinity). Land is black the shaded area marks ocean depths of <3000m. Geostrophic ocean currents in the southern hemisphere are directed so that lower sea level is to the right of the flow direction, which is aligned along lines of equal sea level height. The rise of sea level towards the north defines the eastward flowing Antarctic circumpolar current.

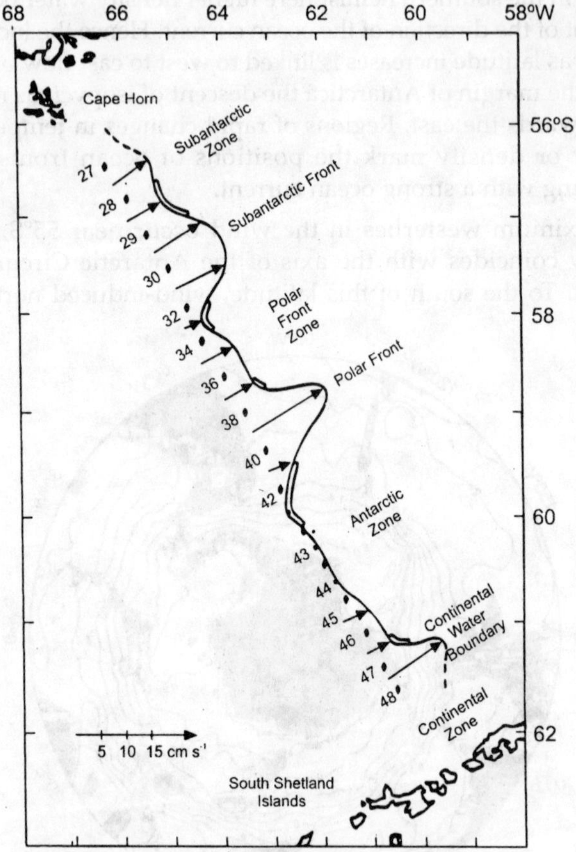

Fig. 12.5 : Vertical averaged geostrophic speed of the upper 2500m directed normal to a line across the Drake Passage. The positions of the front and stratification zones are shown. The highest flows defining the exes of the Antractic Circumpolar Current coincide with the position of the ocean fronts. (Reproduced with permission from Clifford, 1983).

Ekman transport of surface water results in a wide region of upwelling. North of 55°S the Ekman transport diminishes. This causes a region of surface water convergence. Near 55°S surface water sinks under the more buoyant surface water to the north, producing Antarctic Intermediate Water. Antarctic Intermediate Water forms a low salinity layer found at the base of the thermocline of the subtropical southern hemisphere regions. Upwelling poleward of the maximum westerlies brings deeper water to the

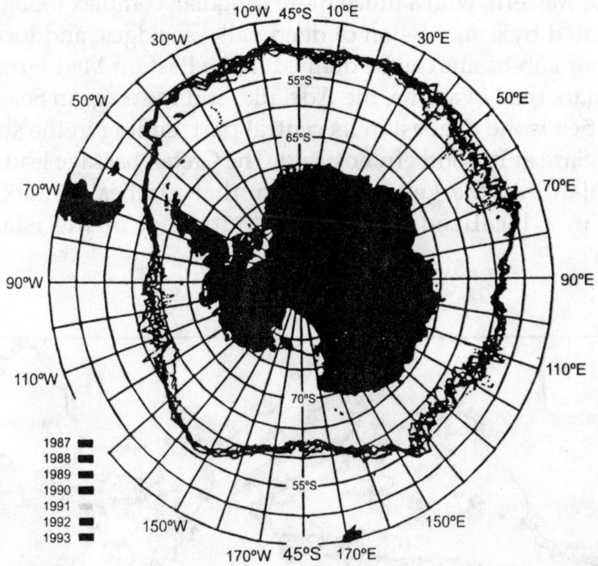

Fig. 12.6 : Position of the Antractic polar front of the year 1987-1993 as revealed by satellite images of sea surface temperature.

sea surface to compensate for the sinking of Antarctic Bottom Water and Antarctic Intermediate Water.

12.4 The Mediterranean Sea Currents

The Mediterranean is a midlatitude semienclosed sea, exchanging water and other properties, with the North Atlantic Ocean.

The Mediterranean Sea is an enclosed basin connected to the Atlantic Ocean by the narrow and shallow Strait of Gibraltar (width ~13km; sill depth ~300m). It is composed of two similar size basins, western and eastern, connected by the Strait of Sicily (width ~35km; sill depth ~250m). The Western Mediterranean has a triangular shape, with the Ligurian Sea at its apex and a large topographic plateau in the Balearic Sea. The islands of Corsica and Sardinia separate the Balearic Sea from the Tyrrhenian Sea, where the bottom relief has the more complex shape of a deep, corrugated valley. The Balearic and Tyrrhenian Seas join in the south in a wide passage between Sardinia and Sicily that leads to the Sicily Straits and the eastern basin.

The Eastern Mediterranean has a more complicated structure than the western, with a much more irregular, complex topography constituted by a succession of deep valleys, ridges, and localized pits. Four sub-basins can be defined in the Eastern Mediterranean: the Ionian, the Levantine, the Adriatic, and the Aegean Seas. The Ionian Sea is the deepest in its central part, ending in the shallow Gulf of Sirte ar its southernmost end. The Cretan passage leads from the Ionian into the Levantine basin, that reaches its maximum depth in a localized depression south-east of the island of Rhodes.

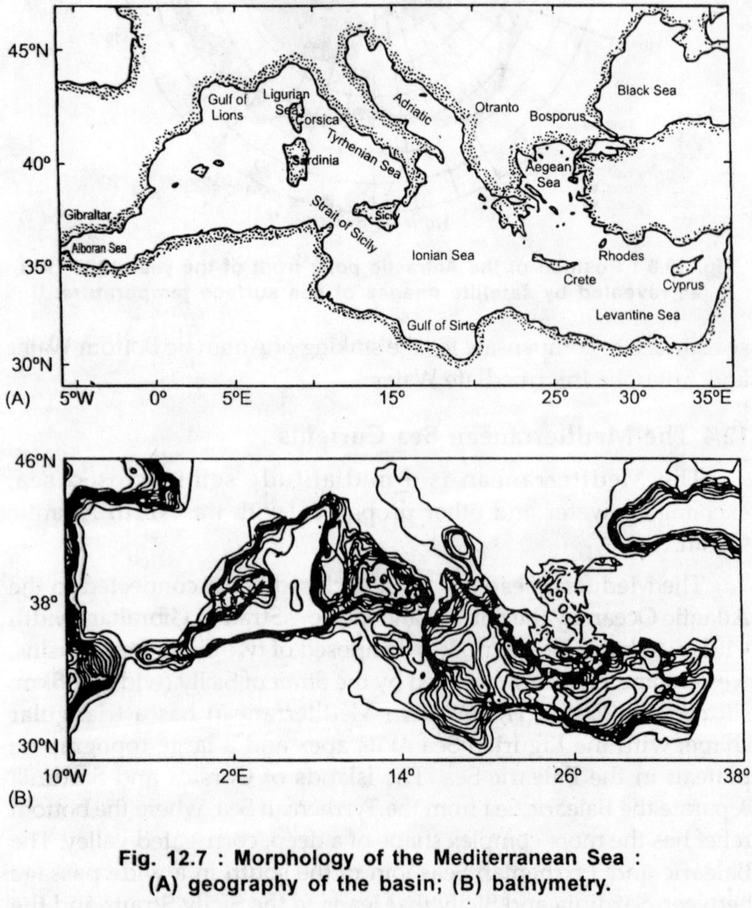

Fig. 12.7 : Morphology of the Mediterranean Sea : (A) geography of the basin; (B) bathymetry.

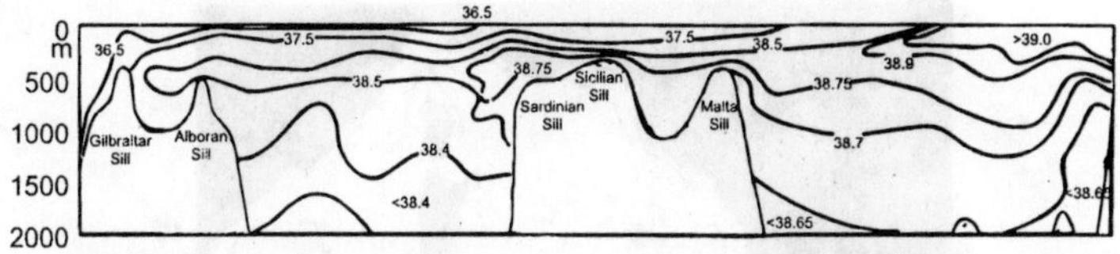

Fig. 12.8 : The open thermohaline cell of the Mediterranean upper layer.

The hydrology and the circulation of the Mediterranean Sea have been known in overall generally for some time. For instance it is well known that the Mediterranean basins, both western and eastern, are evaporation basins (lagoons), with freshwater flux from the Atlantic through the Gibraltar Straits and into the Eastern Mediterranean through the Sicily Straits. The Atlantic Water (AW) mass entering through Gibraltar increases in density because evaporation exceeds precipitation and becomes Modified Atlantic Water (MAW) in its route to the Levantine basin. New water masses are formed here via convection events driven by intense local cooling and evaporation from winter storms. Bottom water is produced in localized convection sites, for the western basin in the Gulf of Lions (Western Mediterranean Deep Water, WMDW) and for the eastern basin in the southern Adriatic (Eastern Mediterranean Deep Water, EMDW), Recent observations also indicate Levantine Deep Water (LDW) formation in the north-eastern Levantine basin during exceptionally cold winters, where Levantine Intermediate Water (LIW) is regularly formed seasonally. Evidence has emerged that LIW formation occurs over much of the Levantine basin but preferentially in the north probably due to meteorological factors. The LIW is the important water mass which circulates westward through both the eastern and western basins and contributes predominantly to the efflux from Gibraltar to the Atlantic, mixed with some EMDW and WMDW.

While the entire Mediterranean is a 'lagoon' for the North Atlantic, the eastern basin itself is a 'lagoon' for the western one. The Atlantic Water (AW) mass enters the Mediterranean at Gibraltar with a typical salinity of S = 36.15 PSU and temperature T = 15°C. The AW becomes Modified Atlantic Water (MAW) through diffusive processes in its pathway eastward and can be identified as a subsurface salinity minimum below ~30 m depth. At the Sicily Straits, the saltier MAW has a salinity S ≤ 37.5 PSU

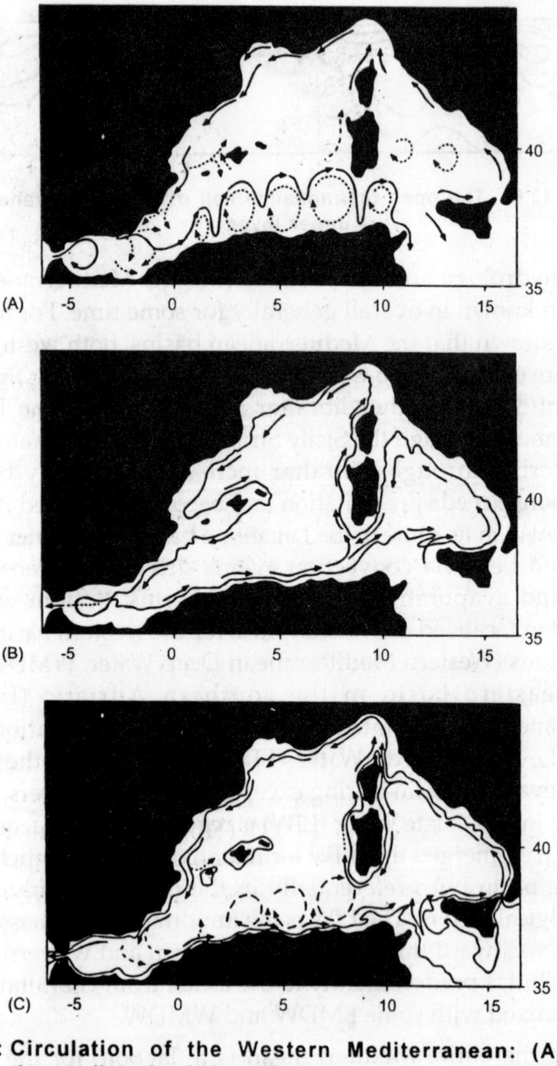

Fig. 12.9 : Circulation of the Western Mediterranean: (A) upper thermocline circulation; (B) intermediate layer circulation with LIW pathways (C) deep thermohaline circulation with WMDW pathways.

reaching a maximum of S < 38.9 PSU in the Levantine basin. Here in its northern part, winter episodes of cold, dry winds blowing from the mainland under surface cooling and evaporative fluxes

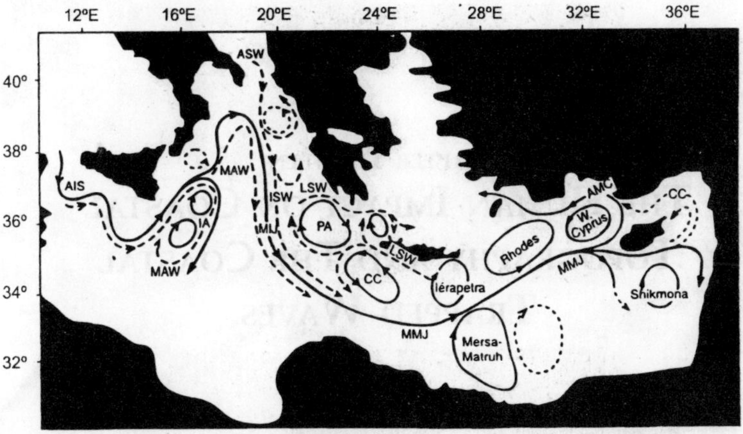

Fig. 12.10 : Eastern Mediterranean: Schematic representation of the upper thermocline circulation, AIS Atlantic Ionian Stream AMC, Asia Minor Current; ASW, Adriatic Surface Water; CC, Cretan Cyclone; IA; Ionian anticyclones; ISW, Ionian Surface Waer; LSW, Levantine Surface Water; MAW, Modified Atlantic Water; MIJ, Mid-Ionian Jet; MMJ, Mid-Mediterranean Jet; PA, Pecas anticyclone.

lead to the formation of LIW, which has $39.0 \leq S \leq 39.2$ PSU and $15°C < \theta < 16°C$ at the formation sites, the most important of which is the well-known Rhodes gyre. The LIW return route is westward in the layer between 200 and 600 m depth. LIW becomes progressively colder and fresher. At the Sicily Straits typical LIW core values are $\theta = 14.3°C$ and $S \leq 38.8$ PSU. At the Gibraltar Straits LIW is diluted to $S \leq 38.5$ PSU, and spreads out in the North Atlantic, becoming Upper North Atlantic Intermediate Water.

Chapter Thirteen

The Human Impact on Coastal Topography and The Coastal Trapped Waves

13.1 Introduction

The major work of humans throughout time is the modification of the landscape. Topography has been modified from the earliest farming to the modern modifications of nature for transportation and commerce (e.g., roads, utilities, mining), and often for recreation, pleasure, and esthetics. While human modifications of the environment have affected vast areas of the continents, and small portions of the ocean floor, nowhere have human intentions met headlong with nature's forces as in the coastal zone.

A most significant change in human behavior since the 1950s has been the dramatic, rapid increase in population and nonessential development in the coastal zone. The associated density of development is in an area that is far more vulnerable and likely to be impacted by natural processes (e.g., wind, waves, storm-surge flooding, and coastal erosion) than most inland areas. Not only are more people and development in harm's way, but the human modifications of the coastal zone (e.g., dune removal) have increased the frequency and severity of the hazards. Finally, coastal engineering as a means to combat coastal erosion and management of waterways, ports, and harbors has had profound and often deleterious effects on coastal environments. The endproduct is a total interruption of sediment interchange berween land and sea, and a heavily modified topography. Natural hazard mitigation is now moving with a more positive, albeit small, approach by restoring narural features, such as beaches and dunes, and their associated interchangeable sediment supply.

The natural coastal zone is highly dynamic, with geomorphic changes occurring over several time scales. Equally significant changes are made by humans. On Ocean Isle, NC, USA, an interior dune ridge, the only one on the island, was removed to make way for development. The lowered elevation put the entire development in a higher hazard zone, with a corresponding greater risk for property damage from flooding and other storm processes.

Another example of change, impacting on property damage risk, can be seen in Kitty Hawk, North Carolina. A large shorefront dune once extended in front of the entire community. The dune was constructed in the 1930s by the Civilian Conservation Corps to halt shoreline erosion, and provide a 'protected' area along which to build a road. The modification was done before barrier island migration was understood. Erosion was assumed to be permanent land loss. The artificial dune actually increased erosion here by acting as a seawall in a long-term sense, blocking overwash sand which would have raised island elevation and brought sand to the backside of the island, although the dune did afford some protection for development. As a consequence, buildings by the hundreds were built in the lee of the dune. Fifty years on, however, the price is being paid. During the 1980s, the dune began to deteriorate due to storm penetration, and the 1991 Halloween northeaster finished the job by creating large gaps in the dune, resulting in flooding of portions of the community. The dune cannot be rebuilt in place because the old dune location is now occupied by the beach, backed up against the frontal road. Between the time of dune construction and 1991, the community had only experienced major flooding once, in the great 1962 Ash Wednesday storm. Between 1991 and 2000, the community was flooded four times.

The effect of shoreline engineering on a whole-island system is starkly portrayed by the contrast between Ocean City, MD and the next island to the south, Assateague Island, MD. It has taken several decades to be fully realized, but the impact of the jetties is now apparent. Assateague Island has moved back one entire island width due to sand trapping by an updrift jetty. Similar stories abound along the coast. The Charleston lighthouse, once on the backside of Morris Island, SC, now stands some 650 m at sea; a sentinel that watched Morris Island rapidly migrate away after the

Charleston Harbor jetties, built in 1898, halted the supply of sand to the island.

Human alterations of the natural environment have direct and indirect effects. Some types of human modifications to the coastal environment include: (1) construction site modification, (2) building and infrastructure construction, (3) hard shoreline stabilization, (4) soft shoreline stabilization, and (5) major coastal engineering construction projects for waterway, port, and harbor management and inlet channel alteration. Each of the modification types impacts the coastal environment in a variety of ways and also has several direct and indirect effects. Some of the effects are obvious and intuitive, but many are surprising in that there can be a domino effect as one simple modification creates potential for damage and destruction by increasing the frequency and intensity of natural hazards at individual sites.

A variety of buildings are constructed in the coastal zone, ranging from single–family homes to high-rise hotels and commercial structures. Some of the common direct effects of building construction are alteration of wind patterns as the buildings themselves interact with natutal wind flow, obstruction of sediment movement, marking the landward limit of the beach or dune, channelizing storm surge and storm-surge ebb flow, and reflection of wave energy. Indirect effects result from the simple fact that once there is construction in an area, people tend to want to add more construction, and to increase and improve infrastructure and services. As buildings become threatened by shoreline erosion, coastal engineering endeavors begin.

Roads, streets, water lines, and other utilities are often laid out in the standard grid pattern used inland, cutting through interior and frontal dunes instead of over and around coastal topography. Buildings block natural sediment flow (e.g., overwash) while the ends of streets and gaps between rigid buildings funnel and concentrate flow, accentuating the erosive power of flood waters. As noted above, during Hurricane Hugo, water, sand, and debris were carried inland along shore-perpendicular roads in several South Carolina communities. Storm-surge ebb along the shore-perpendicular roads caused scour channels, which undermined roadways and damaged adjacent houses and property. Even something as seemingly harmless as buried utilities may cause a problem as the excavation disrupts the substrate, resulting in a less stable topography after post-construction restorations.

Plugging dune gaps can be a part of nourishment and sand conservation projects. Because dunes are critical coastal geomorphic features with respect to property damage mitigation, they are now often protected, right down to vegetation types that are critical to dune growth. Prior to strict coastal-zone management regulations, however, frontal dunes were often excavated for ocean views or building sites, or notched at road termini for beach access. These artificially created dune gaps are exploited by waves and storm-surge, and by storm surge ebb flows. Wherever dune removal for development has occurred, the probability is increased for the likelyhood of complete overwash and possible inlet formation.

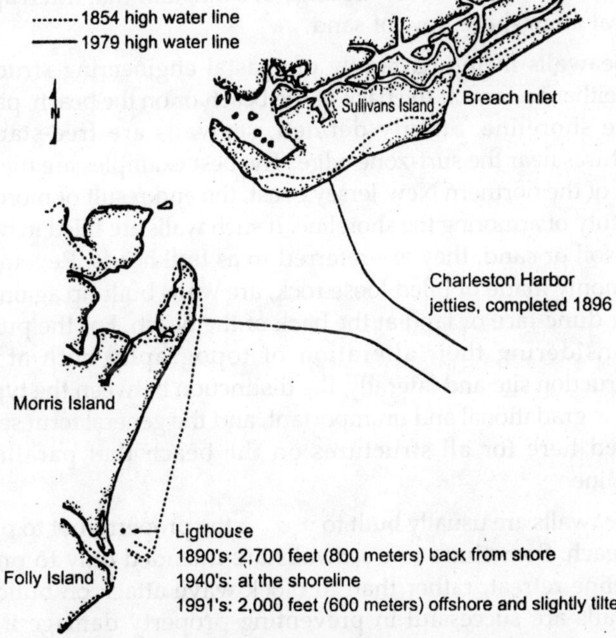

Fig. 13.1 : The jetties that stabilize Charleston Harbor, South Carolina, were completed in 1896. The jetties block the southward transport sediment along the beach. As a result, the islands to the north of the jetties have grown seaward slightly, but the islands to the south of the inlet have eroded back more than 1400m. The Morris Island lighthouse, once on the back side of the island, is now 650m offshore.

13.2 The Importance of Coastal Engineering

Between 80 and 90% of the American open-ocean shoreline is retreating in a landward direction because of sea-level rise and coastal erosion. Because more static buildings are being sited next to this moving and constantly changing coastline, our society faces major problems. Various coastal engineering approaches to dealing with the coastal erosion problem have been developed. More than a centuty of experience with seawalls and other engineering structures in New Jersey and other coastal developments shows that the process of holding the shoreline in place leads to the loss of the beach, dunes, and other coastal landforms. The real societal issue is how to save both buildings and beaches. The action taken often leads to modifications to the coast that limit the natural flexibility of the coastal zone to respond to storms, that inhibit the natural onshore-offshore exchange of sand, and that interrupt the natural alongshore flow of sand.

Seawalls include a family of coastal engineering structures built either on land at the back of the beach or on the beach, parallel to the shoreline. Strictly defined, seawalls are free-standing structures near the surf-zone edge. The best examples are the giant walls of the northern New Jersey coast, the end result of more than a centuty of armoring the shoreline. If such walls are filled in behind with soil or sand, they are referred to as bulkheads. Revetments, commonly made of piled loose rock, are walls built up against the lower dune-face or land at the back of the beach. For the purpose of considering their alteration of topography both at their construction site and laterally, the distinction between the types of walls is gradational and unimportant, and the general term seawall is used here for all structures on the beach that parallel the shoreline.

Seawalls are usually built to protect the property, not to protect the beach. Sometimes low seawalls are intended only to prevent shoreline retreat, rather than to block wave attack on buildings. Seawalls are successful in preventing property damage if built strongly, high enough to avoid being overtopped, and kept in good repair. The problem is that a very high societal price is paid for such protection. That price is the eventual loss of the recreational beach and steepening of the shoreface or outer beach. This is why several states in the USA (e.g. Maine, Rhode Island, North Carolina,

South Carolina, Texas, and Oregon) prohibit or place strict limits on shoreline armoring.

Three mechanisms account for beach degradation by seawalls. Passive loss is the most important. Whatever is causing the shorline to retreat is unaffected by the wall, and the beach eventually retreats up against the wall. Placement loss refers to the emplacement of walls on the beach seaward of the high-tide line, thus removing part or all of the beach when the wall is constructed. Seawall placement was responsible for much of the beach loss in Miami Beach, Florida, necessitating a major beach nourishment project, completed in 1981. Active loss is the least understood of the beach degradation mechanisms. Seawalls are assumed to interact with the surf during storms, which enhances–the rate of beach loss. This interaction can occur in a number of ways including seaward reflection of waves, refraction of waves toward the end of the wall, and intensification of surf-zone currents.

By the year 2000, 50% of the developed shoreline on Florida's western (Gulf of Mexico) coast was armored, the same as the New Jersey coast. Slmilarly 45% of developed shoreline on Florida's eastern (Atlantic Ocean) coast was armored, in contrast to 27% for Sourh Carolina, and only 6% of the developed North Carolina open-ocean shoreline. These figures represent the armored percentage of developed shortlines and do not include protected areas such as parks and National Seashores.

Shoreline stabilization is a difficult political issue because seawalls take as long as five or six decades to destroy beaches, although the usual time range for the beach to be entirely eroded at mid-to-high tide may be only one to three decades. Thus it takes a politician of some foresight to vote for prohibition of armoring. Another issue of political difficulty is that there is no room for compromise. Once a sea-wall is in place, it is rarely removed. The economic reasoning is that the wall must be maintained and even itself protected, so most walls grow higher and longer.

Groins and Jetties

Groins and jetties are walls or barriers built perpendicular to the shoreline. A jetty, often very long (thousands of feet), is intended to keep sand from flowing into a ship channel within an inlet and to reduce the cost of channel maintenance by dredging. Groins are much shorter structures built on straight stretches of beach away from inlets. Groins are intended to trap sand moving in longshore currents. They can be made of wood, stone, concrete, steel, or fabric bags filled with sand. Some designs are referred to as T-groins because the end of the structure terminates in a short shore-parallel segment.

Both groins and jetties are very successful sand traps. If a groin is working correctly, more sand should be piled up on one side of the groin than on the other. The problem with groins is that they trap sand that is flowing to a neighboring beach. Thus, if a groin is growing the topographic beach updrift, it must be causing downdrift beach loss. Per Bruun, past director of the Coastal Engineering program at the University of Florida, has observed that, on a worldwide basis, groins may be a losing proposition, *i.e.*, more beach may be lost than gained by the use of groins. After one groin is built, the increased rate of erosion effect on adjacent beaches has to be addressed. So other groins are constructed, in selfdefense. The result is a series of groins sometimes extending for miles. The resulting groin field is a saw-toothed beach in plan view.

Groins fail when continued erosion at their landward end causes the groin to become detached, allowing water and sand to pass behind the groin. When detachment occurs, beach retreat is renewed and additional alteration of the topography occurs.

Jetties, because of their length, can cause major topographic changes. After jetty emplacement, massive tidal deltas at most barrier island inlents will be dispersed by wave activity. In addition, major build-out of the updrift and retreat of the downdrift shorelines may occur. In the case of the Charleston, SC, jetties noted earlier, beach accretion occurred on the updrift Sullivans Island and Isle of Palms.

Now let us study about offshore breakwaters. Offshore breakwaters are walls built parallel to the shoreline but at some distance offshore, typically a few tens of meters seaward of the normal surf zone. These structures dampen the wave energy on the 'protected' shoreline behind the breakwater, interrupting the longshore current and causing sand to be deposited and a beach to form. Sometimes these deposits will accumulate out to the break-water, creating a fearure like a natural tombolo. As in the case of groins, the sand trapped behind break-waters causes a shortage, sediment downdrift in the directions of dominant longshore transport, leading to additional shoreline retreat (*e.g.*, beach and dune loss, scarping of the fastland, accelerated mass wasting).

13.3 The Impact of Humans on Beaches

In most of the preceding discussion the impact of humans on beaches and shroeline shape and position was emphasized. The beach plays a major role in supplying sand to barrier island and, in fact, is important in supplying sand and gravel to any kind of upland, mainland, or island. In this sense, any topographic modification, however small, that affects the sand supply of the beach will affect the topography. In beach communities, sand is routinely removed from the streets and driveways after storms or when sand deposited by wind has accumulated to an uncomfortable level for the community. This sand would have been part of the island or coastal evolution process. Often, dunes are replaced by flat, well-manicured lawns. Sand-trapping dune vegetation is often removed altogether.

The previously mentioned Civilian Conservation Corps construction of the large dune line along almost the entire length of the Outer Banks of North Carolina is an example of a major topographic modification that had unexpected ramifications, namely the increased rate of erosion on the beach as well as on the backside of the islands. Prior to dune construction, the surf zone, especially during storms, expended its energy across a wide band of island surface which was overwashed several times a year. After construction of the frontal dune, wave energy was expended in a much narrower zone, leading to increased rates of shoreline retreat, and overwash no longer nourished the backside of the island. Now that the frontal dune is deteriorating, North Carolina Highway-12 is buried by overwash sand in a least a dozen places 1–4 times each year. Overwash sand is an important part of the island migration process, because these deposits raise the elevation of islands, and when sediment extends entirely across an island, widening occurs. If not for human activities, much of the Outer Banks would be migrating at this point in time, but because preservation of the highway is deemed essential to connect the eight villages of the southern Outer Banks, the NC Department of Transportation removes sand and places it back on the beach. As a result, the island fails to gain elevation.

Inlet formation also is an important part of barrier island evolution. Each barrier island system is different, but inlets form, evolve, and close in a manner to allow the most efficient means of moving water in and out of estuaries and lagoons. Humans interfere by preventing inlets from forming, by closing them after they open naturally (usually during storms), or by preventing their natural migration by construction of jetties. The net result is clogging of navigation channels by construction of huge tidal deltas and reduced water circulation and exchange between the sea and estuaries.

Globally shoreline change is being affected by human activity that causes subsidence and loss of sand supply. The Mississippi River delta is a classic example. The sediment discharge from the Mississippi River has been substantially teduced by up-stream dam construction on the river and its tributaries. Large flood-control levees constructed along the lower Mississippi River prevent sediment from reaching the marshes and barrier islands along the rim of the delta in the Gulf of Mexico. Natural land subsidence caused by compaction of muds has added to the problem by creating a rapid (1–2 m per centuty) relative sea-level rise. Finally, maintaining the river channel south of New Orleans has extended the river mouth to the edge of the continental shelf, causing most remaining sediment to be deposited in the deep sea rather than on the delta. The end result is an extraordinary loss rate of salt marshes and very rapid island migration. The face of the Mississippi River delta is changing with remarkable rapidity.

Other deltas around the world have similar problems that accelerate changes in the shape of associated marshes and barrier islands. The Niger and Nile deltas have lost a significant part of their sediment supply because of trapping sand behind dams. Land loss on the Nile delta is permanent and not just migration of the outermost barrier islands. On the Niger delta the lost sediment supply is compounded by the subsidence caused by oil, gas, and water extraction. The barrier islands there are rapidly thinning.

Sand mining is a worldwide phenomenon whose quantitative importance is difficult to guage. Mining dunes, beaches, and river mouths for sand has reduced the sand supply to the shoreface, beaches, and barrier islands. In developing countries beachsand mining is ubiquitous, while in developed countries beach and dune mining often is illegal and certainly less extensive, although still a problem. For example, sand mining has adversely affected the beaches of many West Indies nations going through the growing pains of development. Dune mining has been going on for so long that many current residents cannot remember sand dunes ever being present on the beaches, although they must have been there at one time, given the sand supply and the strong winds. For example, on the dual-island nation of Antigua and Barbuda, beach ridges-evidence of accumulating sand – can be observed on Barbuda, but are missing on Antigua. The beach ridges of Barbuda have survived to date only because it is much less heavily developed and populated. Sadly, Barbuda's beach ridges are being actively mined.

Puerto Rico is a heavily developed Caribbean island, much larger than Antigua or Barbuda, and with a more diversified economy. Many of Puerto Rico's dunes have been trucked away. East of the capital city of San Juan, large sand dunes were mined to construct the International Airport at Isla Verde by filling in coastal wetlands. As a result of removing the dunes, the highway was regularly overwashed and flooded during even moderate winter storms. First an attempt was made to rebuild the dune, then a major seawall was built to protect the lone coastal road.

Dredging or pumping sand from offshore seems like a quick and simple solution to replace lost beach sand; however, such operations must be considered with great care. The offshore dredge hole may allow larger waves to attack the adjacent beach. Offshore sand may be finer in grain size, or it may be composed of calcium carbonate, which breaks up quickly under wave abrasion. In all of these cases, the new beach will erode faster than the original beach. Dredging also may create turbidity that can kill bottom organisms. Offshore, protective reefs may be damaged by increased turbidity. Loss of reefs will mean faster beach erosion, as well as the obvious loss to the fishery habitat.

Sand can also be brought in from land sources by dump truck, but this may prove to be more expensive. Sand is a scarce resource, and beaches/dunes have been regarded as a source for mining rather than areas that need artificial replenishment. Past beach and dune mining may well be a principal cause of present beach erosion. In some cases, gravel may be better for nourishment than sand, but the recreational value of beaches declines when gravel is substituted.

Sand mining of beaches and dunes accounts for many of Puerto Rico's problem erosion areas. Such sand removal is now illegal, but permits are given to remove sand for highway construction and emergency repair purposes. However, the extraction limits of such permits are often exceeded–and illegal removal of sand for construction aggregate continues. In all cases, the sand removal eliminates natural shore protection in the area of mining, and robs from the sand budget of downdrift beaches, accelerating erosion. Even a small removal operation can set off a sequence of major shoreline changes.

The Caribe Playa Seabeach Resort along the south-eastern coast illustrates just such a chain reaction. Located west of Cabo Mala Pascua, the resort has lost nearly 15 m of beach in recent years according to the owner. The problem dates to the days before permits and regulation, when an updrift property owner sold beach sand for 50 cents per dump truck load; a bargain by anyone's standards but a swindle to downdrift property owners. Where the sand was removed the beach eroded, resulting in shoreline retreat and tree kills. In an effort to restabilize the shore, and ultimately protect the highway, a rip-rap groin and seawall were constructed. Today, only a narrow gravel beach remains. Undoubtedly much of the aggregate in the concrete making up the buildings lining the shore, and now endangered by beach erosion, was beach sand. What extreme irony: taking sand from the beach to build structures that were subsequently endangered by the loss of beach sand.

The majority of the world's population lives in the coastal zone, and the percentage is growing. As this trend continues, the coastal zone will see increased impact of humans as more loss of habitats, more inlet dredging and jetties, continued sand removal, topography modification for building, sand starvation from groins and jetties, and the increased tourism and industrial use of coasts and estuaries. Our society's history illustrates the impact of humans as geomorphic agents, and nowhere is that fact borne out as it is in the coastal zone. The ultimate irony is that many of the human modifications on coastal topography actually decrease the esthetics of the area or increase the potential hazards.

13.4 The Nature of Coastal–Trapped Waves

Oceanic motion may impinge on the continental shelf. Notably at the equator, waves travel eastward to the coast and divide to travel north and south. In general, oceanic motions accommodate to the presence of the coast and shelf; at a wall, by internal Kelvin waves (vertical structure modes); for more realistic shelf profiles, by coastal trapped waves. Oceanic signals tend to be seen at the coast if along-shelf scale > wave-decay distance or if the feature is shallower than the shelf-water depth.

Natural modes of the ocean are significantly affected by continental shelves. Modes depending on *f/h* gradients have increased frequencies and forms concentrated over topography. Numerical models have shown 13 modes with periods between 30 and 80 hours, each mode being localized over one shelf area.

Atmospheric pressure forcing the sea surface can be effective in driving Kelvin and edge waves, especially if there is some match of speed and scale, more likely in shallower (shelf) seas.

Longshore wind stress τ is believed to be the most effective means of generating coastal trapped waves. Within the forcing region, the flow tends to match the wind field; when or where the forcing ceases, the wave travels onwards ('forwards') and is then most recognizable. A simple view of this forcing is that τ accelerates the alongshore transport *h*u. A more sophisticated view is that τ induces a cross-shelf surface transport $|\tau|\rho f$; coastal blocking induces a compensating return flow beneath, which is acted upon by the Coriolis force to give the same accelerating alongshore transport. A typical stress 0.1 Nm^{-2} for 10^5s (~1 day) accelerates 100m water to 0.1 ms^{-1}. Winds blowing across depth contours may be comparably effective if the coast is distant.

Other generation mechanisms include scattering (of alongshore flow, especially) by shelf irregularities as variable river runoff, and a co-oscillating sea.

Shelf-sea motion is often dominated by tides and responses to wind forcing. On a narrow shelf, oceanic elevation signals penetrate more readily to the coast as the higher-mode decay distances are short; the tide is represented primarily by a Kelvin wave spanning the ocean and shelf. Model fits to semidiurnal measurements, showing a dominant Kelvin wave, have been made off California,

Scotland and north-west Africa, for example. Several areas at higher latitudes show dominant continental shelf wave contributions to diurnal tidal currents, *e.g.*, west of Scotland, Vancouver Island, Yermak Plateau. On a wide shelf, there is correspondingly greater scope for wind-driven elevations and currents. The extensive forcing scale, typically greater than the shelf width, induces flow with minimal structure (typically no reversals across the shelf) corresponding to the lowest-mode continental shelf wave with maximum elevation signal at the coast. In stratified conditions, the upwelling or downwelling response also corresponds to a wave (or waves).

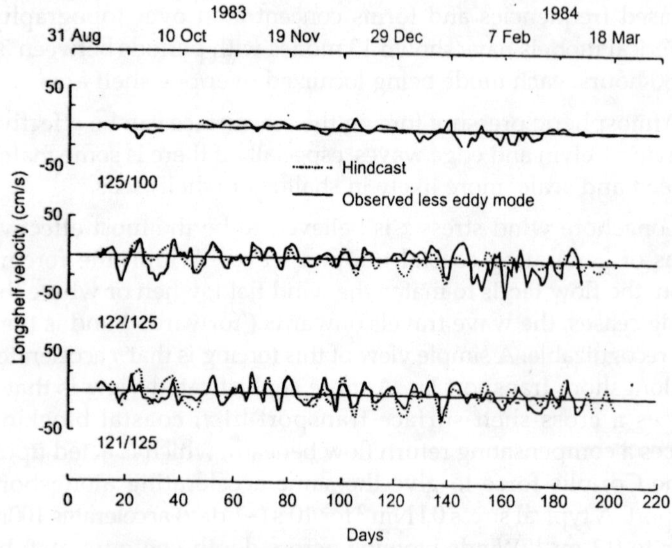

Fig. 13.2 : Comparison of hindcast along-shelf currents using three coastal-trapped wave modes with measured currents a cross-shelf section after band-pass filtering and removing an eddy mode. (Reproduced with permission from Church JA. White Clarke AJ. Freeland HJ and Smith RL (1986). Coastal-trapped waves on the East Australian continental shelf. Part II: model verification. *Journal of Physical Oceanography* 16 : 1945-1957).

As a wave travels, its amplitude is continually incremented by local forcing. A model based on this approach was used to make the hindcast of measured currents on the south-east Australian shelf. At any fixed position, the motion form moves with the local speed $v + [g(h+\eta)]^{1/2}$; crests (η and v positive) gain on troughs (η and v negative) and wave fronts steepen. Similarly, for internal Kelvin waves between an upper layer, depth h, and a deep lower layer (density difference $\Delta\rho$) the local speed is $(gh\Delta\rho/\rho)^{1/2}$ everywhere; troughs gain on crests. Dispersion limits this nonlinear steepening; the associated shorter wavelengths propagate more slowly; and the steepening is left behind by the wave. For small amplitudes and long waves, steepening and dispersion are small and can balance in permanent-form $sech^2(ky+\sigma t)$ solutions. Steepening may be important for internal Kelvin waves, but for typical continental shelf wave amplitudes (near linear) it takes weeks or months, longer than likely frictional decay times.

Mean currents may be forced via frictional contributions to the time-averaged nonlinear convective derivatives in the momentum equations. The mean flows, scale $h^{-1}h'f^{-1}\hat{u}^2$, are typically confined close to the coast or the shelf break (û denotes on–offshore excursion in the waves). Another mechanism is wave-induced form drag over an irregular bottom, giving a biased response to variable forcing; a 'forward' flow along the shelf. For example, low-frequency sinusoidal wind forcing may give a mean current up to a maximum fraction $(2\pi)^{-1}$ of the value under a steady wind, *i.e.*, some centimeters per second, principally in shallower shelf waters.

Three-way interactions between coastal-trapped waves can occur if $\sigma_3=\sigma_1\pm\sigma_2$, $k_3=k_1\pm k_2$, possible for particular combinations according to the shape of the dispersion curve. Typical timescales for energy exchange are many days; effects may be masked by frictional decay. Near a group velocity of zero, there may be more response to a range of energy inputs.

If changes in the stratification and continental shelf form are small in one wavelength, then individual wave modes conserve a longshore energy flux; local wave forms are as for a uniform shelf. Thus Kelvin wave amplitudes increase as $f^{1/2}$ and are confined closer to the coast at higher latitudes. Energy flux conservation implies a large amplitude increase if waves of frequency σ approach a shelf region where the maximum (σ_M) for their particular mode is near σ, as for Scottish and Vancouver Island diurnal tides; the energy tends to 'pile up.' If shelf variations cause σ_M to fall well below σ, then the waves are totally reflected, with large amplitudes near where $\sigma_M = \sigma$. Poleward-propagating waves experience changing conditions. Near the equator, f is small, S (effective stratification) is large and internal Kelvin-like waves are expected, as off Peru; as f increases poleward, waves evolve to less stratified forms, more like continental shelf waves. (Variations of f are special in supporting offshore energy leakage to Rossby waves in the ocean.)

Small irregularities in the shelf (lateral, vertical scales $\varepsilon(gH)^{1/2}/f, \varepsilon H$) generally cause $O(\varepsilon^2)$ effects, but $O(\varepsilon)$ nearby and in phase shifts after depth changes. Scattering occurs, preferentially to adjacent wave modes and (if unstratified) the highest mode at the incident frequency (having near-zero group velocity). However, long waves, L_w, on long topographic variations (as above; L_T) adopt the appropriate local form; scattering is slow unless $L_w \sim L_T$.

If the depth profile has a self-similar form $h[(x - c(y))L(y)]$ then long ($>>L$) continental shelf waves propagate with changes of amplitude but no scattering or change of form, provided that c and L also vary slowly (c/c', $L/L' >> L$). Likewise, there is no scattering if the depth is $h(\xi)$ where $\nabla^2\xi(x,y) = 0$, representing approximately uniform topographic convexity. In these cases, stronger currents and shorter wavelengths are implied on narrow sections of shelf.

Abrupt features are apt to give the strongest scattering, substantial local changes or eddies on the flow. Scattering is the means of slope-current adjustment to a changed depth profile. However, all energy must remain trapped in $\sigma < |f|$; even in $\sigma > |f|$ special interior angles $\pi/(2n + 1)$ can give perfect Kelvin wave energy transmission (for example). Successive reflections in a finite shelf (embayment) may synthesize near-resonant waves with small energy leakage. A complete barrier across the shelf implies reflection into (short, slow) waves of opposite group velocity. There is a considerable literature of particular calculations. However, it is difficult to generalize, because of the several non-scattering cases interspersed among those with strong scattering.

Chapter Fourteen

The CFCs for Ocean Circulation Studies and The Chlorinated Hydrocarbons

14.1 Introduction

The oceans, continents, atmosphere, and cryosphere are part of the climate system. The ocean's role in the climate system involves the transport, sequestration, and exchange of heat, fresh water, and carbon dioxide (CO_2) between the other components of the climate system. When waters descond below the ocean surface they carry with them atmospheric constituents. Some of these are gases such as carbon dioxide and chlorofluorocarbons (CFCs). The CFCs can serve as a physical analog for CO_2 because they are biologically and chemically inert in oceans. In the oceans the distribution of CFCs provides information on which waters have been in contact with the atmosphere in the past few decades. The CFCs also give information on the ocean's circulation and its variability on timescales of months to decades. The timescale information is needed to understand and to assess the ocean's role in climate change, and its capacity to take up anthropogenic constituents from the atmosphere. Thus, the advantage of using tracers like CFCs for ocean circulation studies is the added dimension of time; their time history is fairly well known, they are an integrating quantity and an analog for oceanic anthropogenic CO_2 uptake, and they provide an independent test for time integration of models.

Tracers serve as a 'dye' with which to follow the circulation of ocean waters. There are conventional ocean tracers such as temperature, salinity, oxygen, and nutrients. There are stable isotope tracers such as oxygen-18, carbon-l3, and there are radioactive tracers both naturally occurring (such as the uranium/thorium series, and radium), and those produced both naturally and by the bomb tests (such as tritium and carbon-14). The bomb contributions from the latter two are called transient tracers, as are the CFCs, because they have been in the atmosphere for a short time. This implies an anthropogenic source and a nonsteady input function.

The chlorofluorocarbons, CFCs, are synthetic halogenated methanes. Their chemical structures are as follows: CFC-11 is CCl_3F, CFC-12 is CCl_2F_2, and CFC-113 is CCl_2FCClF_2. For completeness the compound carbon tetrachloride, CCl_4, is also included in this article as its atmospheric source, measurement and oceanic distribution are similar to those of the CFCs. The CFCs have received considerable attention because they are a double-edged environmental sword. They are a threat to the ozone layer, and a greenhouse gas. The CFCs are used as coolants in refrigerators and air conditioners, as propellants in aerosol spray cans, and as foaming agents. These chemicals were developed over 50 years ago when no one realized that they might cause environmental problems. When released CFCs are gases that have two sinks, the predominant one being the atmosphere, and to a lesser extent the oceans. Most of the CFCs go up into the toposphere, where they remain for decades. In the oceans and in the troposphere the CFCs pose no problem. However, some escape into the stratosphere where they are a threat to the ozone layer. Due to their role in UB absorption they have been correlated with the increased incidence of skin cancers. Since the recognition of the CFCs as an environmental problem in the 1970s and the signing of the Montreal Protocol in 1987, the use of CFCs has been phased out. The atmospheric concentrations have just started to decrease. This is an important international step toward correcting the dangerous trend of stratospheric ozone depletion.

The atmospheric CFC concentrations became significant after the 1940s. The concentrations increased exponentially until the mid-1970s, and then increased linearly until the 1990s at a rate of about 5% per year. The production and release data for CFCs tabulated by the Chemical Manufacturers Association (CMA) were used (Figure 1) to reconstruct the atmospheric time histories for the Northern and Southern Hemispheres. Since 1979 the atmospheric concentrations have been based on actual measurements at various sampling stations around the globe, and these are checked against the CMA production and release estimates. The curves in Figure 1 show all CFCs including CCl_4 increasing with time, with CFC-11 leveling off and actually decreasing in the late 1990s. The atmospheric increase of all the CFCs slowed markedly after the Montreal Protocol. The uncertainties in the reconstructed pre-1979 atmospheric histories depend on the atmospheric lifetimes of the compounds. The range of lifertimes for the compounds are 29-76 years for CFC-11, 77-185 years for CFC-12, 54-143 years for CFC-113, and 21-43 years for CCl_4. The uncertainties are a few percent, and they are highest for the early period. The continuous direct atmospheric measurements, which began in the late 1970s, are uncertain to within 1-2%. It is important to evaluate the uncertainties in the atmospheric source function, because they translate into uncertainties when used to put timescales on ocean processes. Because of their long atmospheric residence times, the CFCs are homogeneously distributed in the Northern and Southern Hemisphere, with the Northern Hemisphere about 8% higher than the Southern Hemisphere.

14.2 The Applications of CFCs

Examples of the applications of CFCs to understanding oceanographic processes are divided into four subjects: the thermohaline circulation, upper ocean circulation, model constraints, and biogeochemical processes.

There is a close coupling of the surface waters in high latitudes to the deep ocean through the density-driven thermohaline circulation. During the process of deep-water formation, atmospheric constituents such as CFCs are introduced into the newly formed water. In recent years, major advances in our knowledge of the thermohaline circulation can be attributed to information derived from transient tracer data, particularly for two reasons. First, the development of analytical techniques so that oceanographers can easily produce large quantities of high quality data. Tracer oceanographers have benefited from multi-investigator programs like the World Ocean Circulation Experiment. The following highlights some of the advances that have come about in our understanding of the thermohaline circulation to which observations of CFCs have contributed:

- Discovery of a new water mass component of North Atlantic Deep Water (NADW), called Upper Labrador Sea

Water, location of its formation region and contributing processes, and timescales of eastward spreading along the equator.

- Identification of Denmark Straits Overflow Water as the primaty source of bottom water of the western subpolar basin.
- Confirmation of the structure and continuity of the Deep Western Boundary Current throughout the western North Atlantic Ocean, and extension into the South Atlantic.
- Extension of the CFCs well into the interior of the western North Atlantic show the importance of deep recirculation gyres in ventilating the interior basins, and in slowing the equatorward transport to timescales of < 30 with effective spreading rates of 1-2 cms^{-1}.
- Contribution to quantifying formation rates and decadal climate variability in the Arctic, Greenland and Labrador Seas.
- Estimates for the formation rates of Weddell Sea Deep and Bottom Waters, production rate of Antarctic Bottom Water and pathways and timescales for spreading into the North Atlantic.

The use of CFCs for upper ocean processes has involved the application of concentrations to deduce sources and circulation pathways, and application of pCFC ages. The following highlights some of the advances that have come about in our understanding of the upper ocean circulation to which observations of CFCs have contributed:

- Identification of the Sea of Okhotsk and Alaskan Gyre as important location for the ventilation of North Pacific Intermediate Water, these Waters then spread into the subtropics on a timescale of <20 years.
- Quantification of the flux of water from the mixed layer into thermocline and intermediate layers of the North and South Pacific.

Contribution to the description of sources and pathways of water masses transported from the Pacific through the Indonesian Seas into the Indian Ocean.

- Quantification of the sources of northern and southern water and the processes needed to ventilate the tropical Pacific and Atlantic, including advection, diapycnal and vertical mixing.
- Observation that pathways of the most recently ventilated Antarctic Intermediate Waters are into the eastern South Indian Ocean, while at that level there appears to be flow of older waters from the South Pacific into the western Indian Ocean.
- Quantification of subduction and formation rates for subtropical underwaters and in the North Atlantic its interannual variability that is negatively correlated with intermediate waters of the eastern subpolar gyre.

In general CFC concentrations and inventories have been used in comparison with model simulated concentrations and inventories. The time-dependent nature of the CFCs provides a stringent test of a model's ability to integrate property distributions over time. The following highlights some of the advances that have come about in our ability to put constraints on models from the use of CFCs in models:

- Dilution of CFCs transported by the Deep Western Boundary Current and effect on tracer ages.
- Testing the sensitivity of a model for correct simulation of formation rates, pathways, and spreading rates.
- Testing the sensitivity of a model for correct simulation of ocean model velocity fields.
- Determining the model sensitivity to subgrid scale mixing for purposes of estimating ventilation rates.
- The importance of considering seasonal variations in the upper oceans as part of the tracer boundary conditions when trying to simulate subduction processes.
- Demonstration in a model simulation that eddy transport is required to transport South Indian subtropical gyre waters across the equator along the western boundary.
- Use of CFCs to validate model parameterizations of gas fluxes.

The tracers provide a method for calculating rates of biogeochemical fluxes that is independent of direct biological measurements. Again the age information from the CFCs is used to calculate rates for these processes. (*see* Nitrogen Isotopes in the Ocean). The following highlights some of the advances that have come about in our understanding of biogeochemical processes to which observations of CFCs have contributed:

- Apparent oxygen utilization rates from the central Arctic that are so high, they need to be balanced by transport of high production water from over the continental shelves.
- Quantification of moderate biological consumption and initially low oxygen concentrations in the Arabian Sea are needed to maintain the low oxygen layer.
- Calculation of denitrification rates for the Arabian Sea and Bay of Bengal.

The advantage of oceanic tracers like CFCs is that they can be used to provide timescale information for oceanographic processes. Direct application of the timescale information from the CFCs is used to calculate fluxes of atmospheric constituents, such as CO_2. The oceans have taken up a considerable portion of the anthropogenic CO_2 released to the atmosphere. A large part of the uptake involves water mass formation in high latitudes. The rate at which these waters are transported into the interior will have an effect on the rate at which anthropogenic CO_2 is taken up.

14.3 The Presence of DDT and PCBs in Marine Environment

Analyses during the early 1970s of various species of marine biota for DDT family chemicals and PCBs established widespread distribution of these chemicals in many areas of the world's oceans from equatorial to polar regions and to depths of 4000-5000m in the Atlantic Ocean. Biota with significant lipid (fat) in their body tissues tended to accumulate DDT family compounds and PCBs much more than biota with lower lipid content. Marine mammals and birds accumulated higher concentrations of DDT compounds and PCBs, presumably due to being near the top of the food web (biomagnification of contaminants), and having a high body lipid content (especially for marine mammals and birds).

Analyses of surface seawater samples (both dissolved and particulate) and samples of the atmosphere over the oceans established the presence of low concentrations (ranges of 0.01-1 $ngkg^{-1}$ water or $0.001ngm^{-3}$ of air) of DDT family compounds and PCBs. Very low concentrations and difficulties in avoiding contamination from the sampling ship and the sampling gear made measurement of deep seawater samples problematic. The few deep-water samples analyzed documented that the concentrations of DDT and PCBs were not higher than about 0.001 $ngkg^{-1}$ of water. Confirmatory measurements were not made for years thereafter because of intense debate among chemical oceanographers about how to make reliable measurements for these compounds at very low concentrations in sea water. In contrast, the underlying sediments had accumulated sufficient concentrations, because of sorption on particles and deposition, to allow undisputed measurements of both DDT and PCB in deep ocean surface sediments.

Several surveys and research programs documented much higher concentrations of DDT and PCBs in coastal waters and coastal ecosystems compared to open ocean ecosystems; especially near.

Recent progress with measuring low concentrations of PCBs in deep ocean waters has enabled a few measurements of deep ocean waters. A depth profile of the sum of concentrations of several chlorobiphenyl congeners for a station in the eastern North Atlantic documents higher concentrations in surface warers with decreasing concentrations with depth as expected due to the greater contact of the surface waters with the atmosphere and contemporary environment. However, all the concentrations are very low in comparison to concentrations found in near-shore waters, lakes, and rivers. Concentrations in the deepest waters are below or at detection limits of the analytical methods used. PCBs in mid-depth and deeper waters are most likely a result of the flux of particles from the surface water carrying sorbed PCBs in and on particles to deeper waters. There has been considerable progress during the 1980s and 1990s in understanding the role and details of particles as conveyers of chemicals from the surface ocean to deep waters and sediments. As the particles sink through the deep waters, desorption and disaggregation of particles and subsequent desorption releases PCBs. Therefore, it is likely, though not proved due to lack of a series of data over time, that current deep water concentrations of PCBs reflect inputs from particle fluxes over the total time of PCB use and release to the environment; for example inputs from peak use in North America and Europe during the 1950s to 1960s.

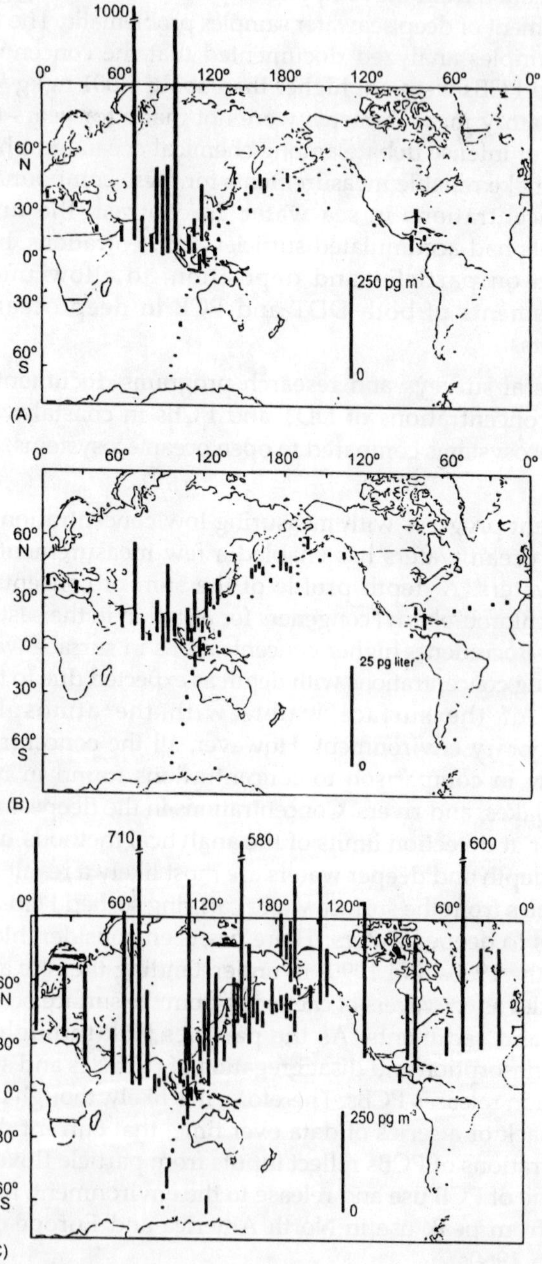
1000
0° 60° 120° 180° 120° 60° 0°
60° N
30°
0°
30°
60° S
250 pg m⁻³
0
(A)
0° 60° 120° 180° 120° 60° 0°
60° N
30°
0°
30°
60° S
25 pg liter⁻¹
0
(B)
710
580
600
0° 60° 120° 180° 120° 60° 0°
60° N
30°
0°
30°
60° S
250 pg m⁻³
0
(C)

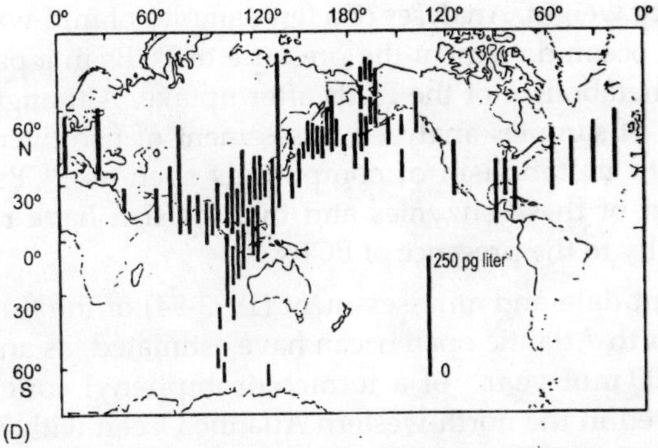

Fig. 14.1 : Concentrations of DDT (A, B) and PCSs (C,D) in the atmosphere and surface sea water (B,D) 1989-90. (Adapted with permission from Iwata *et al.*, 1993)

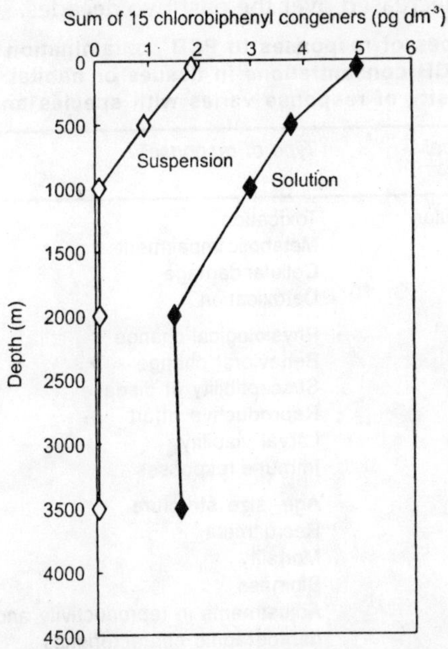

Fig. 14.2 : Depth profile of PCB concentrations in sea water, May 1992 at 47°N, 20°W. Adapted from Petrick *et al.*, 1996.

Deep-water sediments contain low concentrations of DDT and PCBs in the range of 1-100 ngg^{-1} dry weight or parts per billion. Relatively few deep-ocean benthic (bottom dwelling) animals have been analyzed, but those that have been analyzed contain detectable concentrations of DDT and PCBs in the range of

1 μgg^{-1} dry weight. Analyses of a few samples of mid-water fish in the deep ocean document the presence of PCBs in a pattern that reflects metabolism of the PCBs after uptake. Although small in numbers of samples analyzed, assessment of fish liver enzymes specific to metabolism of compounds such as PCBs indicate activation of these enzymes and that the fish have responded biologically to the presence of PCBs.

Recent data and an assessment (1993-94) of the flux of PCBs to the North Atlantic open ocean have estimated, as an example, that 14000 mol year^{-1} of a tetrachlorobiphenyl congener was sequestered in the north-western Atlantic Ocean with 75% of the flux being in the deep-ocean pelagic sector and only 25% in the coastal-continental shelf area. This same assessment provided good reasoning that the flux of PCBs from the atmosphere to the North Atlantic has decreased over the past two decades.

General types of responses to PCB contamination for marine organisms. (PCB concentrations in tissues or habitat types eliciting a given intensity of response varies with species and ecosystem)

Level of biological organization	*Type of response*[a]
Biochemical-cellular	Toxication Metabolic impairment Cellular damage Detoxication
Organismal	Physiological change Behavioral change Susceptibility of disease Reproductive effort Larval viability Immune responses
Population	Age, size structure Recruitment Mortality Biomass Adjustments in reproductivity and other demographic characteristics
Community	Species abundance Species distribution Biomass Trophic interactions

Coastal Ocean Aquatic toxicological studies during the 1960s to the present have established the range of sensitivities of several different species and life stages of marine organisms to chlorinated hydrocarbon pesticides and industrial chemicals such as PCBs. These studies established lethal tolerance limits for these chemicals in aquatic habitats, usually in the microgram per liter range. Such concentrations are rarely encountered today in estuarine or coastal areas, except in the immediate vicinity of exceptional discharges of the chemical.

Even though the chlorinated hydrocarbons are among the chemicals more resistant to chemical or biological alteration in the environment, there are physical-chemical (e.g., sorption-desorption, transfer from water to air), microbial transformation and degradation, and animal enzyme modifications or transformations, that change the mixture of compounds as the chemicals move through the environment. For example, the mixture of chlorobiphenyl congeners found in a lobster were dramaticallv different from the original mixture discharged in a nearby effluent from a capacitor manufacturing facility. Those chlorobiphenyl congener mixtures were also different in composition compared to the PCB congener mixture in flounder caught in the same area. The sediment congener mixture for the habitat of both the lobster and flounder had yet another composition. Biological effects of chlorinated hydrocarbons can be specific to each individual chemical in terms of mode of action and potency of action. Thus, the presence of complex and diverse mixtures of these chemicals in various compartments of an ecosystem introduces significant complications to the task of providing an assessment of ecological and human health risks associated with a given site of chlorinated hydrocarbon contamination. In addition, the presence of other chemicals of

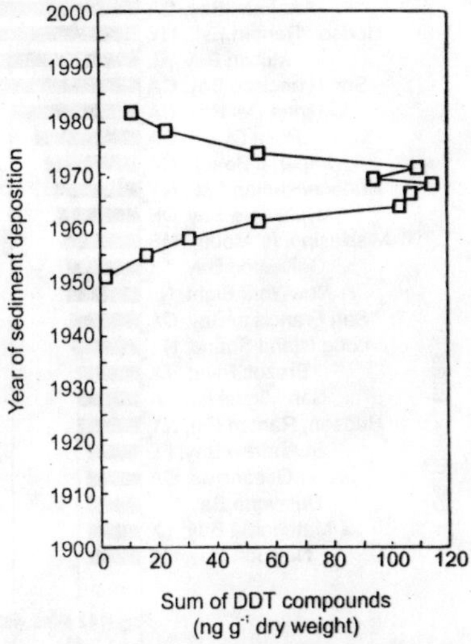

Fig. 14.3 : Depth profile of DDT concentrations in San Petro Shelf sediments, Southern California, USA, documenting the historical input of DDT to the area. (Adapted from Eganhouse and Kaplan 1988.)

environmental concern in many of the same areas, means that the present knowledge base is insufficient to provide a high degree of accuracy to a quantitative risk assessment for ecological and human health concerns.

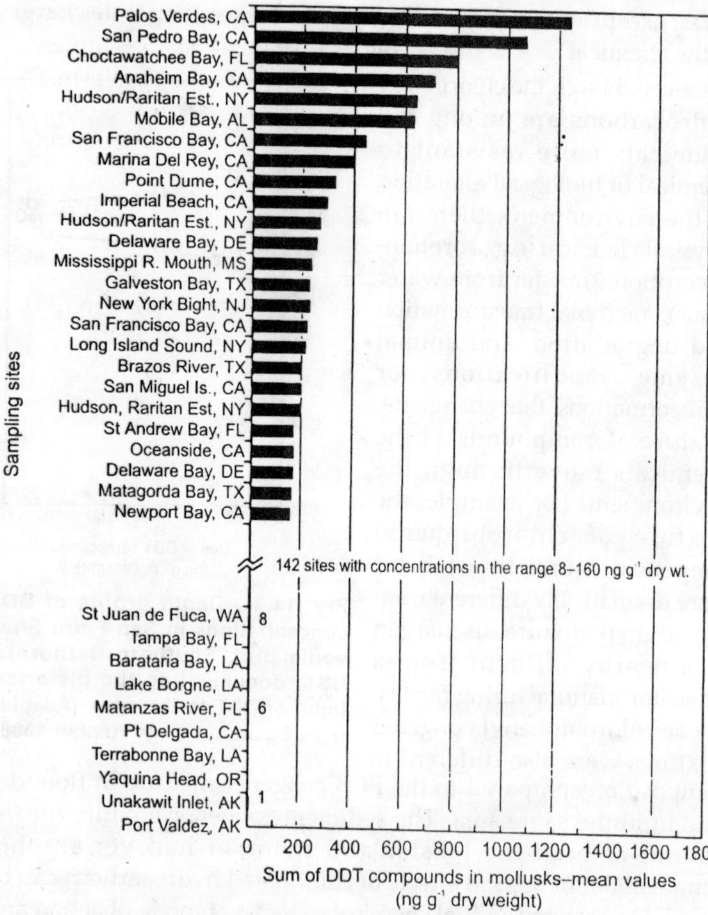

Fig. 14.4 : Concentration of (A) DDT and (B) PCBs in Mussel Watch Stations US NOAA Status and Trends Program (Musseis and Oysters) 1986. (Adapted from *A Summary of Data on Tissue Contamination from the First Three Years (1986–1988) of the Mussel Watch Project*. NOAA Technical Memorandum NOS OMA 49. NOAA Office of Oceanography and Marine Assesment, Ocean Assessments Division, Rockville, MD US, 1989).

Humanity was fortunate to learn valuable lessons from early results that indicated widespread regional and global transport of chlorinated pesticides and PCBs accompanied by environmental and human health problems. Much knowledge has been gained about the biogeochemical cycles and about ecological and biological effects of these chemicals in the oceans. This knowledge has been used to guide policy and management actions in many instances. Otherwise, the deplorable situation faced by Inuits might be much more severe and widespread and natural resource populations, including oceanic species, might have been more severely impacted.

REFERENCES

American Steel and Wire Co. *Wire Rope Engineering Handbook.* American Steel and Wire Co., Div. U.S. Steel, New Jersey, 1946, 140 pp.

An Experiment's 45-day Undersea Saturation Dive at 205 Feet. *Sealab II* Project Group, eds. D. C. Pauli and G. P. Clapper, ONR Report ACR-124, Office of Naval Research, Washington, D. C., 1967.

Behnke, A. R., R. M. Thompson, and E. P. Motely. "Physiologic effects from breathing air at four atmospheres," *Am. J. Physiology,* 112 (1935), 554-558.

Brauer, R. W. Personal Communications of *Ocean Science News,* 10, No.4, (January 26, 1968), p. l.

British Ropes, Ltd. Results of Repeated Loading Tests on Viking Nylon Ropes. Personal communication from British Ropes Ltd., Leith, Edinburgh, Scotland, 1950.

Brown, J. L. "Orientation to the Vertical during water immersion," *AERO SP. MED.,* 32 (1961), 209–217.

Columbian Rope Co. *Columbian Ropes for Every Use.* Catalog No. 53, Columbian Rope Co., Auburn, N. Y., 1954, 68 pp.

Columbian Rope Co. Test Data Sheets. Personal communication from Columbian Rope Co., Auburn, N.Y., Aug. 1960.

Galerne, A. (personal communication, 1968) Mr. Galerne has stated, from the evidence gathered during 1,100-foot dry chamber dives on helium-oxygen mixtures, that commercially available helium descramblers work satisfactorily. This may be the technical breakthrough anticipated.

(a) Gerschmann, R. "Biological Effects of Oxygen," pp. 475–494; (b) Haugaard, N. "The Toxic Action of Oxygen on Metabolism and the Role of Trace Metals," pp. 495–507. From *Oxygen in the Animal Organism.* New York: The MacMillan Company, 1964.

Gillen, H. W. "Oxygen Convulsions in Man," *Third International Conference on Hyperbaric Medicine,* National Academy of Sciences, Publication 1404. ed. I. W. Brown, Jr. and B. G. Cox. Washington, D.C., 1966, pp. 217–222.

Gillen, H. W. "Saturation Exposures: Advantages and Limitations," Chapter 11, *Topics in Ocean Engineering, Houston:* The Gulf Publishing Company, 1969.

Grover, H. J., S. A. Gordon, and L. R. Jackson. *Fatigue of Metals and Structures.* NAVAER 00-25-534, Bureau of Aeronautics, U.S. Navy, Washington, D.C., 1954.

Joosting, W. C. Q. "Investigation into Long Period Waves in Ports," *Proc. XIX International Navigation Congress* (London, 1957), Sect. II, Communication I, pp. 205–227.

Judge, A. W. *Engineering Materials: V. III Theory and Testing of Materials.* London: Sir Isaac Pitman & Sons, Ltd., 1930, Chap. VII, pp.174–210.

Kaplan, P. and Putz, R. R. "The Motions of a Moored Construction Type Barge in Irregular Waves and Their Influence on Construction Operation," Tech. Report (Contract NBy 32206, Marine Advisers, Inc.), U.S. Naval Civil Eng. Lab., Port Hueneme, Calif., Aug. 1962.

Kennedy, A. J. *Processes of Creep and Fatigue in Metals.* London: John Wiley & Sons, Inc., 1962, pp. 234-285.

Kent, D.R. and S. Weissman. "Visual Resolution Underwater." Report No. 476. U.S. Naval Submarine Medical Center, Groton, Connecticut, May 5, 1966.

Kilner, F. A. "Model Tests on the Motion of Moored Ships Placed on Long Waves," *Proc. 7th Conf. Coastal Eng.*, The Hague, Netherlands, (Aug. 1960), Council Wave Research, Univ. Calif., Berkeley, 1961, pp. 723–745.

Kinney, Jo Ann S., S. M. Luria, S. Weissman, and H. H. Matteson. Underwater Visibility of Fluorescent and Non-Fluorescent Paints." Memorandum Report No. 65-11, U.S. Naval Submarine Medical Center, Groton, Connecticut, September 1, 1965.

Lanphier, E. H. "Nitrogen-oxygen mixture physiology. Added respiratory a dead space." (Value in personnel selection tests.) U.S. Navy Experimental Diving Unit. Project NS 185-005. Subtask No.5, Test No.4, Washington, D.C., 1956.

Leendertse, J. J. "Analysis of the Response of Moored Surface and Subsurface Vessels to Ocean Waves," Tech. Report RM-3368-PR, The Rand Corp., Santa Monica, Calif., 1963.

Leendertse, J. J. "Analysis of the Response of Offshore-Moored Ships to Waves," *Proc. 9th Conf. Coastal Eng.* Lisbon, Portugal, (June, 1964), *ASCE Coastal Eng. Research Council*, New York, (1965), pp. 733–755.

Luria, S. M., Jo Ann A. Kinney, and S. Weissman. "Estimation of Size and Distance Underwater." Report No. 462. U.S. Naval Submarine Medical Center, Groton, Connecticut, December 10, 1965.

MacInnis, J. B. "The Medical and Human Performance Problems of Living under the Sea," *Canad. Med. A.J.*, 95 (1966), 191–200.

Miskelly, R. E. "Nylon and Other Synthetic Fiber Ropes," Conf. Fire Dept. Instructors, Memphis, Tenn., Jan. 1947.

Moseley, O. H. "Tensile Strength of Manila Rope," Selected Eng. Paper No. 88, Inst. C. E. (London), 1930, 6 pp.

O'Brien, J. T. and Kuchenreuther, D. I. (1957) "Forces Induced on a Large Vessel by Surge," Proc. *ASCE (Journal, Waterways & Harbors Div.)*, Paper 1571, (March 1958), 29 Pp.

O'Brien, J. T. and Kuchenreuther, D. I. (1959) "Forces Induced on a Large Vessel by Surge" (closure discussion), *Proc. Inst. C. E., 125*(I) (1960), Pp. 883–888.

Plymouth Cordage Company. *Plymouth Marine Ropes; Their Construction; Their Uses.* Plymouth Cordage Co., Plymouth, Mass., c. 1960.

Plymouth Cordage Company. Test Data Sheets. Personal communication from Plymouth Cordage Co., Plymouth, Mass., Aug. 1960.

Porter Company. *How to be a Wire Rope Expert.* H. K. Porter, Co., Inc., Leschen Wire Rope Div., St. Louis, Mo., c. 1960, 15 Pp.

Putz, R. R. "The Motions of Ship-Shaped Vessels in Surface Wave Environments," *Proc. Offshore Exploration Conf.*, Long Beach (Feb. 1966), M. J. Richardson, Inc., pp. 297-312.

Russell, R, C. H. "A Study of the Movement of Ships Subjected to Wave Action," *Proc. Inst. C. E. (London)*, v. 12 (April 1959), p.379. Discussions, *Proc. ICE*, 15 (April 1960), Pp. 435–450.

Schaefer, K. E. "Effects of prolonged diving training," In *Second symposium on Underwater Physiology*. ed. C. J. Lambertsen and L. J. Greenbaum, Jr., National Research Council. Publication 1181, Washington, D.C., 1963, pp. 271–278.

Sergeant, R. L. "Speech during respiration of a mixture of helium and oxygen," *Aerospace Med.*, 34 (1963), 826–829.

Sommet, J. "The Motion of a Ship Under the Action of Seiche," *Proc. Sympos. Behavior of Ships in a Seaway*, Wageningen, Netherlands, (1957), 1, pp. 354–373.

Stewart, D. M. "Behavior of Stationary Wire Ropes in Tension and Bending," *Trans. ASCE, 102*, (1937), 606-650.

Wilson, B. W. "A Case of Critical Surging of a Moored Ship," *Proc. ASCE*, 85 (WW4) (Dec. 1959), 159-176, also (with discussion) *Trans. ASCE*, 126 (1961), 571–603.

Wilson, B. W. "Characteristics of Anchor Cables in Uniform Ocean Currents," Tech. Rept. No. 204–1 (Ref. 60-5T), Texas A & M Research Foundation, College Station, Texas, April 1960, 154 pp.

Wilson, B. W. "Characteristics of Anchor Cables in Uniform Ocean Currents," *Bulln. Permanent International Assoc. Navig. Congr.*, Nos. 11, 12, 1964.

Wilson, B. W. "Research and Model Studies on Range Action in Table Bay Harbour," Cape Town, Doctoral thesis to Univ. Cape Town, South Africa, 1951, v.I, 463 pp.; v. II, 216 figs (unpublished).

Wilson, B. W. "Ship Response to Range Action in Harbor Basins," *Proc. ASCE*, 76 (Separate No. 41), (Nov. 1950), 29 pp.; also *Trans. ASCE*, 116, (1951), pp. 1129-1157.

Wilson, B. W. "The Energy Problem in the Mooring of Ships Exposed to Waves," Proc. Princeton Univ. Conf. on "Berthing and Cargo Handling in Exposed Locations," Princeton, N. J., Oct. 1958; also *Bulln. Inrernational Navig. Congr.*, No. 50, 1959, 65 pp.

Withy, M. O. and Washa, G.W. *Materials of Construction*. New York: John Wiley & Sons, Inc., 1954, Chap. 26, pp. 10-13.

(a) Zetterstrom, A. "Deep-Sea Diving with Synthetic Gas Mixtures," *Military Surgeon*, 103 (1948), 104–106. (b) Bjurstedt, H. and G. Severin. "The Prevention of Decompression Sickness and Nitrogen Narcosis by the Use of Hydrogen as a Substitute for Nitrogen (The Arne Zetterstrom Method for Deep-Sea Diving) *Military Surgeon*, 103 (1948), 107–116.

Index